De Volson Wood

Turbines, Theoretical and Practical

With Numerical Examples and Experimental Results and Many Illustrations. Second

Edition

De Volson Wood

Turbines, Theoretical and Practical
With Numerical Examples and Experimental Results and Many Illustrations. Second Edition

ISBN/EAN: 9783337106218

Printed in Europe, USA, Canada, Australia, Japan

Cover: Foto ©berggeist007 / pixelio.de

More available books at **www.hansebooks.com**

TEXT BOOKS

FOR

ENGINEERS AND STUDENTS.

BY

DE VOLSON WOOD,

Professor of Engineering in Stevens Institute of Technology.

A TREATISE ON **THE** RESISTANCE OF **MATERIALS, AND**
AN APPENDIX **ON** THE PRESERVATION OF TIMBER.
Seventh edition. **8vo,** cloth $2 00

This work originally consisted chiefly of the Lectures delivered by Professor Wood
upon this subject to the classes in Civil Engineering in the University of Michigan, but it
has since been twice revised. It is a text-book, and gives a brief sketch of the history of
the development of the theories connected with the growth of the subject, and a large
amount of experimental matter. An English reviewer of the work says : " It is equal in
grade to Rankine's work."

A TREATISE ON THE **THEORY** OF THE **CONSTRUCTION**
OF **BRIDGES** AND **ROOFS.** Designed as a **Text-Book and for**
Practical Use. Illustrated with numerous wood-engravings. **Sixth**
edition. 1 vol., 8vo $2 00

The development of the subject in this work is progressive in its character. It begins
with the most elementary methods, and ends with the most general equations. The skel-
eton forms of the triangular trusses, including the Warren girder; and quadrangular forms,
including the Long, Pratt, Howe, Towne, Whipple, Post, and other forms. The examples
are original and many are novel in character.

THE **ELEMENTS OF ANALYTICAL MECHANICS.** With numerous examples and illustrations. For use in Scientific Schools and
Colleges. Seventh edition, including Fluids. 8vo, cloth $3 00

The Calculus and analytical methods are freely used in this work. It contains many
problems completely solved, and many others which are left as exercises for the student.
The last chapter shows how to reduce all the equations of mechanics from the principle of
D'Alembert.

PRINCIPLES OF **ELEMENTARY MECHANICS.** Fully illustrated.
Ninth edition. 12mo, cloth $1 25

The chief aim of this work **is to define,** explain and enforce the fundamental principles
of mechanics. The analysis is simple, but the work is comprehensive. The principles of
energy are applied to the **solution of** many problems. There is a variety of examples at the
end of each chapter. A novel feature of the work is the "Exercises," which contain many
peculiar and interesting **questions.**

SUPPLEMENT AND KEY TO **PRINCIPLES OF ELEMENTARY**
MECHANICS. 12mo, cloth $1 25

This work not only contains a solution of all the examples and answers to the exercises,
but much additional matter which can hardly fail to interest any student of this science.

THE **ELEMENTS** OF CO-ORDINATE GEOMETRY. In **Three**
Parts. I. Cartesian Geometry and Higher Logic. II. Quarternions.
III. **Modern** Geometry and an Appendix. 1 vol., 8vo, cloth. New
edition, **with** additions. $2 00

It was designed to make this a thoroughly practical book for class use. The more
abstruse parts of the subject are omitted The properties of the conic sections are, so far
as practicable, treated under common heads, thereby enabling the author to condense the
work. The most elementary principles only of quarternions and modern geometry are
presented ; but in these parts, as well as in the first part, are numerous examples.

A **TREATISE ON** CIVIL ENGINEERING. By Prof. D. H. Mahan.
Revised and edited, with additions and new plates, by Prof. De Volson
Wood. With an Appendix and Complete Index. 8vo, cloth. . .$5 00

TRIGONOMETRY. Analytical, Plane, and **Spherical.** With Logarithmic
Tables. Fourth edition. By Prof. De Volson **Wood.** 12mo, cloth $1 00

THERMODYNAMICS. By Prof. De Volson Wood, C.E., M.E. 12mo,
cloth. Second edition $4 00

THEORY OF TURBINES. By Prof. De Volson Wood. 8vo, cloth, $2 50

TURBINES,

THEORETICAL AND PRACTICAL,

WITH NUMERICAL EXAMPLES AND EXPERIMENTAL RESULTS
AND MANY ILLUSTRATIONS.

DE VOLSON WOOD,

Professor of Mechanical Engineering, Stevens Institute of Technology.

SECOND EDITION,
REVISED AND ENLARGED.

NEW YORK:
JOHN WILEY & SONS.

LONDON
CHAPMAN & HALL, LIMITED.

—

1896.

REMARK.

About sixty-five years ago M. Poncelet made a solution of the Fourneyron turbine which, for its thoroughness and the directness of its analysis, has become classical (*Comptes Rendus*, 1838). But that writer neglected the frictional (and other) resistances within the wheel, and assumed that the buckets, or passages in the wheel, were constantly full. The former is an important element in the theory, and its consideration makes the analysis but little more complicated.

Weisbach, in his *Hydraulic Motors*, gives a solution in which frictional resistances are involved, and the sections of the stream at the outlet of the supply chamber, the entrance into the wheel, and all the sections of the buckets are determined when the wheel runs for best efficiency. The formulas, however, are so complex that but little practical knowledge can be gained from their general discussion. I have, therefore, assumed that the wheels here discussed have about the proportions made for commercial purposes, and deduced certain numerical results which are entered in tables ; and a simple examination of these furnishes certain desirable information.

The driving power here considered is that of an incompressible fluid, which in practice will be water. The steam turbine or those driven by a compressible fluid are not in practice constructed like water turbines, and no theory for such is here attempted.

The first part of this work, to page 65, develops the general theory of Turbines, and the latter part treats of actual wheels, their forms, construction, capacity and efficiency.

THE AUTHOR.

HYDRAULIC MOTORS.

A general solution of all classes of turbines, including the frictional resistances, is here attempted. There are two general classes: one in which the water enters the buckets freely, with a velocity due to the head, and the other in which its entrance is more or less resisted by the pressure in the buckets, and hence the velocity is less than that due to the head. The former are called turbines of " free deviation ;" the latter, " pressure" turbines.

There will first be given a *general solution* of the "pressure turbine," and the other turbines will be considered as special cases of the more general one.

2. *Notation.*

Let Q be the volume of water passing through the wheel per second,

δ, the weight of unity of volume of the water, or 62½ pounds per cubic foot; then

$\delta Q = W$ will be the weight of water passing through the wheel per second,

h_1 be the head in the supply chamber above the entrance to the buckets,

h_2, the head in the tail race above the exit from the bucket,

z_1, the fall in passing through the buckets,

$H = h_1 + z_1 - h_2$, the effective head,

U, the useful work done by the water upon the wheel,

R, the work lost by frictional resistances, whirls, etc.,

μ_1, the coefficient of resistance along the guides,

μ_2, the coefficient of resistance along the buckets,

r_1, the radius of the initial rim,

r_2, the radius of the terminal rim,

ρ, the radius to any point of the bucket,

$n = r_1 \div r_2$, the ratio of the initial to that of the terminal radius,

V, the velocity of the water issuing from supply chamber,

v_1, the initial velocity of the water in the bucket in reference to the bucket,

v, the velocity along the bucket at any point,

v_2, the terminal velocity in the bucket,

V_n, the velocity of exit in reference to the earth,

ω, the angular velocity of the wheel,

α, terminal angle between the guide and initial rim $= CAB$,

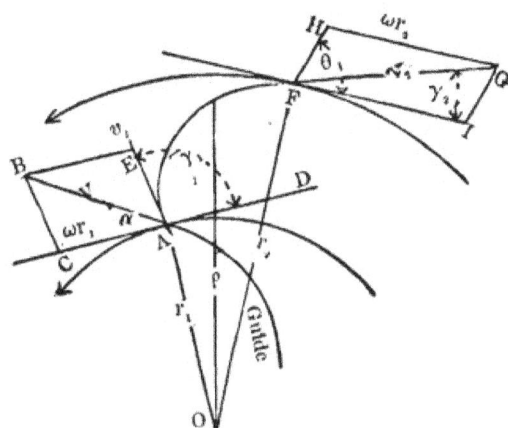

Fig. 1.

γ_1, angle between the initial element of bucket and initial rim $= EAD$,

$\gamma_2 = GFI$, the angle between the terminal rim and terminal element of the bucket,

$\theta = HFI$, angle between the terminal rim and actual direction of the water at exit,

p_1, the pressure of water at entrance of the bucket per unit area,

p, the pressure of water at any point of the bucket,

p_s, the pressure of water at exit,

p_a, the pressure of an atmosphere,

$a = eb$, the arc subtending one gate opening, Fig. 3,

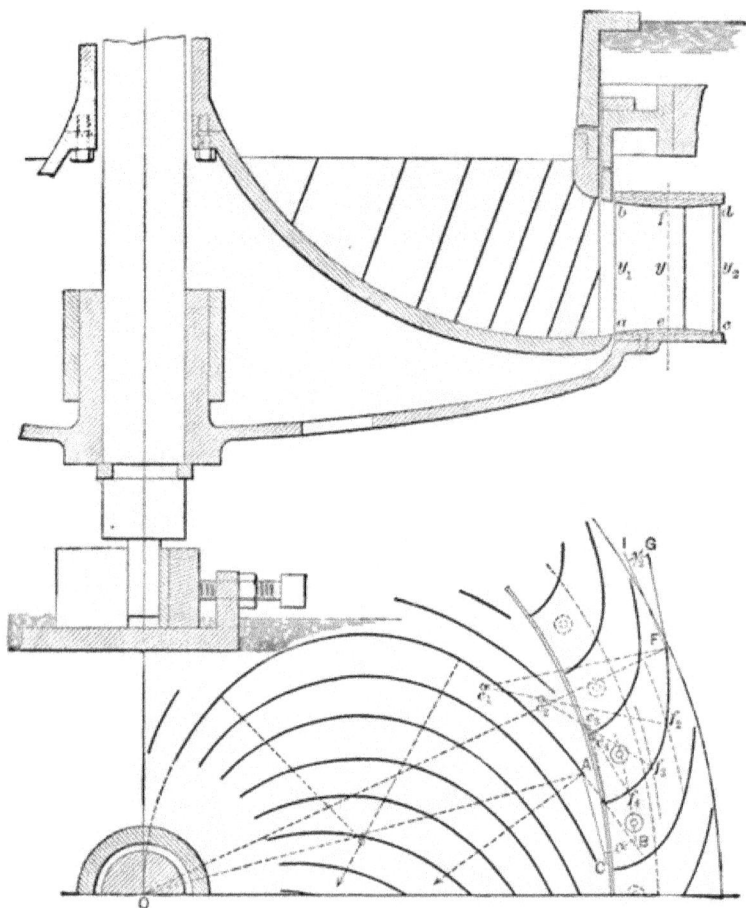

RADIALLY OUTFLOW TURBINE.

a_{3}, the arc subtending one bucket at entrance. (In Fig. 3, a and a_{1} appear to be the same, but in practice they are usually different, a being greater than a_{1}.)

$a_{2} = gh$, the arc subtending one bucket at exit.

K, normal section of passage, bf being its projection, it being assumed that the passages and buckets are *very* narrow ;

k_{1}, initial normal section of bucket, bd its projection ;

k_{2}, terminal normal section, gi being its projection ;

Y, the depth of K, y_{1} of k_{1}, and y_{2} of k.

Then

$$K = Ya \sin a ;$$
$$k_{1} = y_{1}a_{1} \sin y_{1} ; \quad (1)$$
$$k_{2} = y_{2}a_{2} \sin \gamma_{2}, $$

$\omega r_{1} = $ velocity of initial rim,

$\omega r_{2} = $ velocity of terminal rim.

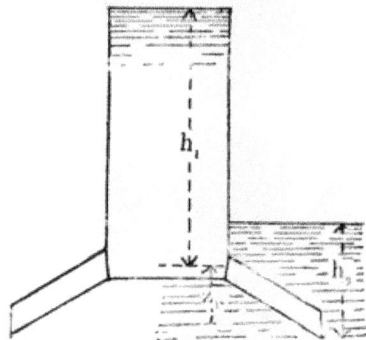

FIG. 2.

3. *General Solution.*

Beginning with the pressure on the top of the supply chamber, the relation between the heads, actual and virtual, will be determined to the point of discharge from the wheel.

The pressure per unit on the upper surface of the supply chamber will be that of the atmosphere, or

$$p_{a},$$

and the corresponding virtual head in terms of a column of water will be

$$\frac{p_{a}}{\delta}.$$

The head in the supply chamber above the entrance to the wheel will be

$$h_{1} ;$$

therefore, the total head above the initial element of the bucket will be

$$h_1 + \frac{p_a}{\delta}.$$

This head produces an actual pressure p_1 at the entrance to the bucket and the velocity V of exit from the guides; hence, according to Bernoulli's theorem, the heads due to the pressure p_1 and velocity V, will equal the former, or

$$h_1 + \frac{p_a}{\delta} = \frac{p_1}{\delta} + \frac{V^2}{2g}; \quad \cdots \quad (2)$$

$$\therefore p_1 = p_a + \delta h_1 - \delta \frac{V^2}{2g}, \quad (3)$$

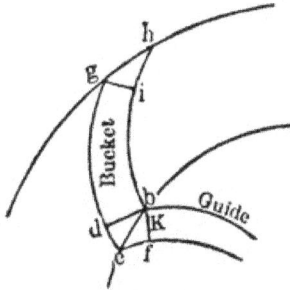

FIG. 3.

which will be the theoretical pressure at entrance to the bucket if friction be neglected. Represent the head lost by friction by

$$\mu_1 \frac{V^2}{2g},$$

which must also be overcome by the head in the supply chamber, so that we have, by adding it to the second member of (2) and transforming,

$$(1 + \mu_1) V^2 = 2g h_1 + 2g \left(\frac{p_a - p_1}{\delta} \right). \quad \cdots \quad (4)$$

The triangle of velocities ABC, Fig. 1, gives

$$V = \frac{\sin \gamma_1}{\sin (\alpha + \gamma_1)} \omega r_1 = \frac{\sin \gamma_1}{\sin (\alpha + \gamma_1)} n \omega r_2 \quad \cdots \quad (5)$$

$$v_1 = \frac{\sin \alpha}{\sin (\alpha + \gamma_1)} \omega r_1. \quad \cdots \quad (6)$$

The relation between the initial and terminal velocities in

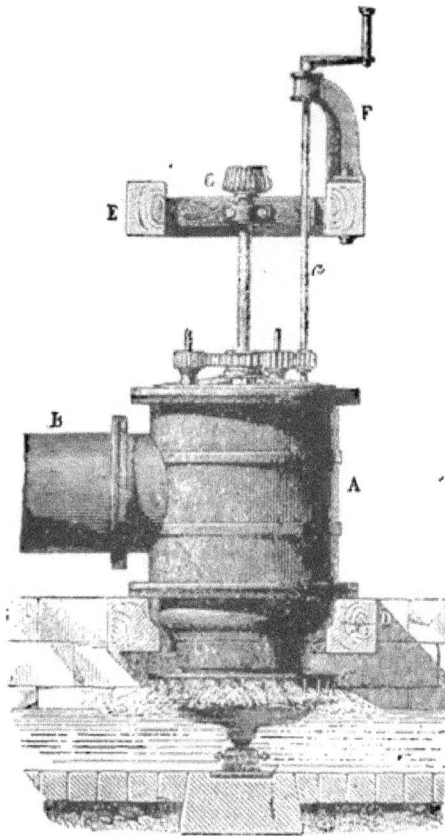

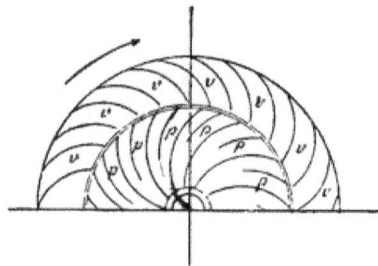

FOURNEYRON TURBINE

the bucket involves the velocity of the wheel and the pressure in the bucket.

Let m be an elementary mass at a distance ρ from the axis of the wheel, then will the centrifugal force be

$$m \omega^2 \rho,$$

and if this element by moving a distance ds in the tube also moves outward a distance $d\rho$, the work done by the centrifugal force will be

$$m \omega^2 \rho \, d\rho.$$

If the tube (or bucket) be inclined downward, the work done (or energy acquired) by the weight in falling a distance dz will be

$$m g dz.$$

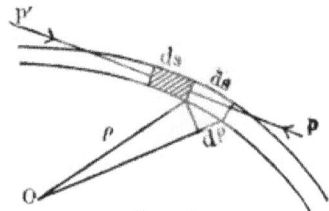

FIG. 4.

These two works will be expended in the following ways:

a. Increasing the energy of the water in the tube in reference to the tube by an amount

$$\tfrac{1}{2} \, m d \, (v^2).$$

b. In doing work against the difference of pressures on the two faces of the element, and considering the back pressure p greater than the forward pressure p', the work will be

$$m g \frac{dp}{\delta},$$

where $dp \div \delta$ is equivalent to a head through which mg would work.

c. In overcoming frictional resistance. The law of frictional resistances is not well known, but is assumed to vary as the energy of the mass and wetted perimeter. The perimeter is here discarded, hence the work will be

$$\mu_2 \cdot \tfrac{1}{2} mv^2 ds.$$

Hence we have

$$mgdz + m\omega^2\rho d\rho = \tfrac{1}{2} md(v^2) + mg\frac{dp}{\delta} + \tfrac{1}{2}\mu_2 mv^2 ds. \quad . \quad . \quad (7)$$

But the last term cannot be integrated unless v be a known function of s, and since this is not known, we make $v = v_2$, the terminal velocity. The coefficient μ_2 is determined independently of the length, and includes the value $\mu_2 s$, when s is the length of a bucket.

Integrating between initial and terminal limits gives

$$z_1 + \frac{\omega^2(r_2^2 - r_1^2)}{2g} = \frac{v_2^2 - v_1^2}{2g} + \frac{p_2 - p_1}{\delta} + \mu_2\frac{v_2^2}{2g}. \quad . \quad (8)$$

The fall z_1 is so small in practice compared with the next term of the equation, that it may, and will, be omitted, giving

$$(1 + \mu_2)v_2^2 = v_1^2 + \omega^2(r_2^2 - r_1^2) - 2g\frac{p_2 - p_1}{\delta}, \quad . \quad . \quad (9)$$

which gives v_2.

At exit the pressure will be

$$p_2 = p_a + \delta h_2. \quad . \quad . \quad . \quad . \quad . \quad . \quad (10)$$

The velocity of exit, relative to the earth, will be

$$V_2^2 = v_2^2 + \omega^2 r_2^2 - 2v_2.\omega r_2 \cos\gamma_2. \quad . \quad . \quad . \quad (11)$$

The work done upon the wheel will be the initial (potential) energy of the water less the energy in the water as it quits the wheel, still further diminished by the energy due to frictional losses; or

$$U = \delta QH - \delta Q\frac{V_2^2}{2g} - R. \quad . \quad . \quad . \quad . \quad (12)$$

$$R = \mu_1\delta Q\frac{V^2}{2g} + \mu_2\delta Q\frac{v_2^2}{2g}. \quad . \quad . \quad (13)$$

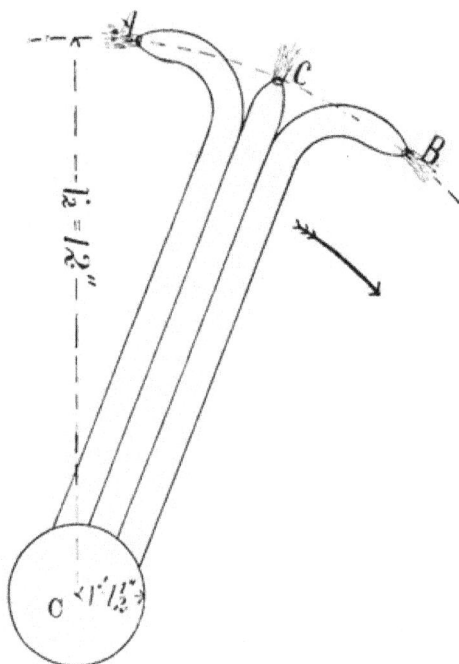

Three tubes having small orifices, A, B, C, on the same arc, rotate about a common axis, o, one discharges to the left, another radially, and the third to the right; they receive water at the ends near to the axis, with so small a velocity as to be negligible; required the velocity of exit. Equation (9) gives for all three cases $v_2 = \omega^2 r_2^2$ for the velocity relative to the tubes: and equation (11) gives for 1st, $V_2 = 2\,\omega\,r_2$

"   "   2d, $V_2 = \sqrt{2}\,.\,\omega\,r$

"   "   3d, $V_2 = 0$.

A sufficient number of equations have now been established to find, by elimination, the useful work U in terms of the angular velocity ω and known constants; and the efficiency will be U divided by the *theoretical* work the water was capable of doing. Performing the operations, there will be found

$$E = \frac{U}{\delta Q H} =$$

$$\frac{r_2 \omega}{g H \sqrt{1 + \mu_2}} \times \left\{ \sqrt{1 + \mu_2} \left\{ -1 + \frac{\cos \alpha \sin \gamma_1}{\sin(\alpha + \gamma_1)} \left(\frac{r_1}{r_2}\right)^2 \right\} r_2 \omega + \cos \gamma_2 \times \right.$$

$$\left. \sqrt{2gH + \left\{ 1 - \left(\frac{2 \cos \alpha \sin \gamma_1}{\sin(\alpha + \gamma_1)} + \mu_1 \frac{\sin^2 \gamma_1}{\sin^2(\alpha + \gamma_1)} \left(\frac{r_1}{r_2}\right)^2 \right) \right\} r_2^2 \omega^2} \right\} \quad . \quad (14)^*$$

*To deduce (14).

$$\frac{2gU}{\delta Q} = 2gH - V_2{}^2 - \mu_1 V^2 - \mu_2 \sigma_2{}^2. \text{ from (13 and 12)} \qquad = 2gH - (1 + \mu_2)\sigma_2{}^2 - \omega^2 r_2{}^2 + 2\omega r_2 \cos \gamma_2.\sigma_2 - \mu_1 V^2 \text{ from (11)}$$

$$= 2gH - \sigma_1{}^2 - 2\omega^2 r_2{}^2 + \omega^2 r_1{}^2 + 2g \frac{p_a - p_1}{\delta} + 2gh_2. - \mu_1 V^2 + \frac{2\omega r_2 \cos \gamma_2}{\sqrt{1 + \mu_2}} \sqrt{v_1{}^2 + \omega^2 (r_2{}^2 - r_1{}^2) - 2g \frac{p_a - p_1}{\delta} - 2gh_2}$$

(by means of 9 and 10)

From (4).

$$2g\frac{p_a - p_1}{\delta} = (1 + \mu_1) V^2 - 2gh_1.$$

Substituting this, and σ_1 from equation (6) and V' from (5) gives

$$\frac{2gU}{\delta Q} = -2\omega^2 r_2{}^2 + \omega^2 r_1{}^2 - 2gH + \omega^2 (r_2{}^2 - r_1{}^2) + 2gH + \left(\frac{\sin^2 \alpha}{\sin^2(\alpha + \gamma_1)} - \frac{\sin^2 \gamma_1}{\sin^2(\alpha + \gamma_1)} \right) \omega^2 r_1{}^2 -$$

$$+ \frac{2\omega r_2 \cos \gamma_2}{\sqrt{1 + \mu_2}} \sqrt{\omega^2 (r_2{}^2 - r_1{}^2) + 2gH + \left(\frac{\sin^2 \alpha}{\sin^2(\alpha + \gamma_1)} - \frac{\sin^2 \gamma_1}{\sin^2(\alpha + \gamma_1)} \right) \omega^2 r_1{}^2 - \mu_1 \frac{\sin^2 \gamma_1}{\sin^2(\alpha + \gamma_1)} \omega^2 r_1{}^2.}$$

$$= L\left[- M^2 r_2 \omega + \cos \gamma_2 \sqrt{2gH + N^2 r_2^2 \omega^2}\right] r_2 \omega, \quad \ldots \quad (15)$$

when

$$\left.\begin{array}{l}
L = \dfrac{1}{gH\sqrt{1 + \mu_2}}. \\[2mm]
-M^2 = \sqrt{1 + \mu_2}\left[- 1 + \dfrac{\cos \alpha \sin \gamma_1}{\sin (\alpha + \gamma_1)}\left(\dfrac{r_1}{r_2}\right)^2\right], \\[2mm]
N^2 = 1 - \left(\dfrac{2 \cos \alpha \sin \gamma_1}{\sin (\alpha + \gamma_1)} + \mu_1 \dfrac{\sin^2 \gamma_1}{\sin^2 (\alpha + \gamma_1)}\right)\left(\dfrac{r_1}{r_2}\right)^2
\end{array}\right\} \quad (15a)$$

For maximum efficiency make $dE \div d\omega = 0$ in (15) and solve for ω, calling this particular value ω', then

$$\omega' = \frac{\sqrt{gH}}{r_2}\sqrt{\frac{M^2 - \sqrt{M^4 - N^2 \cos^2 \gamma_2}}{N^2 \sqrt{M^4 - N^2 \cos^2 \gamma_2}}}, \quad \ldots \quad (16)$$

which value substituted in equation (15) will give the maximum efficiency. Then equations (5), (6), become

$$V = \frac{\sin \gamma_1}{\sin (\alpha + \gamma_1)} \omega' r_1, \quad \ldots \ldots \quad (17)$$

$$v_1 = \frac{\sin \alpha}{\sin (\alpha + \gamma_1)} \omega' r_1. \quad \ldots \ldots \quad (18)$$

Also from (9), (10), (4), (17), (18),

But

$$\frac{\sin^2 \alpha - \sin^2 \gamma_1}{\sin^2 (\alpha + \gamma_1)} = \frac{\sin^2 \alpha - \sin^2 \gamma_1}{\sin^2 \alpha - \sin^2 \gamma_1 + 2 \sin (\alpha + \gamma_1) \cos \alpha \sin \gamma_1}$$

$$= 1 - \frac{2 \sin (\alpha + \gamma_1) \cos \alpha \sin \gamma_1}{\sin (\alpha + \gamma_1) \sin (\alpha - \gamma_1) + 2 \sin (\alpha + \gamma_1) \cos \alpha \sin \gamma_1}$$

$$= 1 - \frac{2 \cos \alpha \sin \gamma_1}{\sin (\alpha + \gamma_1)},$$

which substituted above will give equation (14).

Substituting eq. (16) in (15) gives, by reduction,

$$E_{max.} = \frac{1}{N^2 \sqrt{1+\mu_2}}[M^2 - \sqrt{M^4 - N^2\cos^2\gamma_2}] \qquad (16a)$$

$$\therefore \omega^2 = \frac{gH\sqrt{1+\mu_2}}{r_2^2\sqrt{M^4 - N^2\cos^2\gamma_2}} E_{max.} \quad \cdots \quad (16b)$$

$$v_2 = \frac{1}{\sqrt{1+\mu_2}}\sqrt{2gH + N^2\omega'^2 r_2^2}, \quad \cdots \quad (19a)$$

If the terminal angle of the guide, α, the initial and terminal angles of the bucket γ₁ and γ₂, respectively, the ratio of the radii r₁, r₂ and the frictional resistances constant; then for all such wheels M and N will be constant and the efficiency will be constant; the velocity of the initial rim, ω′ r₁, the velocity through the gate V, the initial velocity in the bucket, will each and all vary as $\sqrt{H}$.

$$v_2 = \sqrt{\frac{1}{1 + \mu_2}}$$

$$\times \sqrt{2gH + \left(r_2{}^2 - 2\frac{\cos\alpha\sin\gamma_1}{\sin(\alpha+\gamma_1)}r_1{}^2 - \mu_1\frac{\sin^2\gamma_1}{\sin^2(\alpha+\gamma_1)}r^2_1\right)\omega^2} \quad . \quad (19)$$

The normal sections of the buckets will be

$$K = \frac{Q}{V}; \quad k_1 = \frac{Q}{v_1}; \quad k_2 = \frac{Q}{v_2}; \quad k = \frac{Q}{v}. \quad . \quad . \quad (20)$$

The depths of those sections will be

$$Y = \frac{K}{A\sin\alpha}, \quad y_1 = \frac{k_1}{a_1\sin\gamma_1}, \quad y_2 = \frac{k_2}{a_2\sin\gamma_2}, \quad . \quad (21)$$

DISCUSSION.

4. *Three simple systems are recognized.*

$r_1 < r_2$, called *outward flow.*

$r_1 > r_2$, called *inward flow.*

$r_1 = r_2$, called *parallel flow.*

The first and second may be combined with the third, making a *mixed* system. The third, in theory, is really an inward or outward flow, with an indefinitely narrow crown, although the analysis applies to a parallel flow wheel, in which the width is indefinitely small, and depth small compared with the total head.

5. *Value of γ_2, the quitting angle.*

Equations (14) and (16a) show that the efficiency is increased as $\cos\gamma_2$ is increased, or as γ_2 decreases, and is greatest for $\gamma_2 = 0$. Hence, theoretically, the terminal element of the bucket should be tangent to the quitting rim for best efficiency. This, however, for the discharge of a finite quantity of water, would require an infinite depth of bucket, as shown by the third of equations (21). In practice, therefore, this angle must have a finite value. The larger the diameter of the terminal rim the smaller may be this angle for a given depth of wheel and given quantity of water discharged. Theoretical considerations then would require, for best efficiency, a very large diameter for the quitting rim, and a very small angle, γ_2, between the terminal element of the bucket and the rim; but commercial considerations require some sacrifice of best efficiency to cost, so that a smaller diameter and larger angle of discharge is made. If wheels are of the same diameter and depth, the inward flow wheel requires a larger quitting angle for the same volume of water than the outward flow, since the discharge rim will be smaller in the former than in the latter wheel, and the velocity v_2, eq. (19), will also be less. In practice γ_2 is from $10°$ to $20°$.

6. *Relation between γ_2 and ω'.*

Equation (16) when put under the form

$$\omega' = \frac{\sqrt{gH}}{Nr_2}\sqrt{\frac{1}{\sqrt{1 - \frac{N^2\cos^2\gamma_2}{M^4}}} - 1} \quad . \quad . \quad (22)$$

shows that ω' increases as γ_2 decreases, and is largest for $\gamma_2 = 0$; that is, in a wheel in which all the elements except γ_2 are fixed, *the velocity of the wheel for best effect must increase as the quitting angle of the bucket decreases.*

If the terminal element be radial, then $\gamma_2 = 90°$, and equation (22) appears to give $\omega' = 0$; but the discussion really fails for this case. See article 89.

7. *Values of $\alpha + \gamma_1$.*

If $\alpha + \gamma_1 = 180°$, and α and γ_1 both finite, then will M and N in (15a) both be infinite; but equation (5) gives

$$\omega = \frac{\sin 180°}{\sin \gamma_1} \cdot \frac{V}{r_1} = 0 ; \quad . \quad . \quad \quad . \quad . \quad . \quad (23)$$

that is, the wheel will have no motion, and no work will be

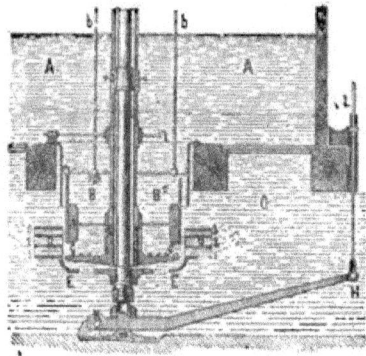

FOURNEYRON TRIPLE WHEEL.

done. If $\alpha + \gamma_1 = 180°$, then the terminal element of the guide and the initial element of the bucket have a common tangent. in which case the stream can flow smoothly from the former into the latter only when the wheel is at rest. (See Fig. 5.)

If $\alpha + \gamma_1$ exceed 180°, ω' would be negative, and it would be necessary to rotate the wheel backwards in order that the water should flow smoothly from the guide into the bucket.

It follows, then, that $\alpha + \gamma_1$ must be less than 180°, but the best relation cannot be determined by analysis ; however, since the water should be deflected from its course as much as possible from its entering to its leaving the wheel, the angle α for this reason should be as small as practicable.

8. Values of α.

If $\alpha = 0$, equation (14) will reduce to

$$E = \frac{\omega}{gH\sqrt{1 + \mu_2}}\left[\sqrt{1 + \mu_2}\,(- r_2^2 + r_1^2)\,\omega \right.$$
$$\left. + r_2 \cos \gamma_2 \sqrt{2gH + (r_2^2 - 2r_1^2 - \mu_1 r_1^2)\,\omega^2} \right] \quad (24)$$

which is independent of γ_1 ; hence, for this limiting case, the efficiency will be independent of the initial angle of the bucket. This is because the water enters the wheel tangentially and therefore has no radial component that would give an initial velocity in the bucket ; and equation (18) shows that the initial velocity v_1 would be zero, while (17) shows that the velocity of the initial rim must equal that of the water flowing from the guides, or
$$V = \omega' r_1.$$

For the limiting, or critical case,
$$\alpha = 0, \; \gamma_2 = 0, \; u_1 = 0, \; \mu_2 = 0,$$
the velocity producing maximum efficiency will be, from equation (16),
$$r_1 \omega' = \sqrt{gH.} \qquad . \qquad (25)$$

or the velocity of the initial rim, if the wheel be frictionless, will be that due to half the head in the supply chamber.

If $r_2^2 = 2r_1^2$, then

$$r_2\omega' = \sqrt{2gH}, \quad \ldots \quad (26)$$

or the velocity of the terminal rim will equal that due to the head. Substituting in (19) the values $\alpha = 0$, $\gamma_2 = 0$, $\mu_1 = 0$, $\mu_2 = 0$, $r_2^2 = 2r_1^2$, and it will reduce to

$$v_2 = \sqrt{2gH}, \quad . \quad . \quad . \quad . \quad (27)$$

as it should.

The following table gives the values of quantities for the three classes of wheels:

TABLE I.

	$\alpha = 0,$		$\gamma_2 = 0,$		$\mu_1 = 0.$		$\mu_2 = 0.$	
DIMENSIONS OF WHEEL.	VELOCITY OF		Velocity of Exit from Guide V.	VELOCITY IN BUCKET.		Velocity of Exit w.	Efficiency. E.
	Inner Rim.	Outer Rim.		Initial v_1.	Terminal v_2.		
	ωr_1	ωr_2					
$r_1 = \sqrt{\tfrac{1}{2}}r_2$	$\sqrt{gH}$	$\sqrt{2gH}$	$\sqrt{gH}$	0.00	$\sqrt{2gH}$	0.00	1.000
$r_1 = r_2$	$\sqrt{gH}$	$\sqrt{gH}$	$\sqrt{gH}$	0.00	$\sqrt{gH}$	0.00	1.000
	ωr_2	ωr_1					
$r_1 = 1.4 r_2$	$0.714\sqrt{gH}$	$\sqrt{gH}$	$\sqrt{gH}$	0.00	$.714\sqrt{gH}$	0.00	1.000

In the first case the inner rim is the initial one, in the third case the outer rim is initial, it being an inward flow wheel.

Since, in this case, the velocity of admission to the wheel in reference to the earth is that due to half the head in the supply chamber, and the velocity of exit is zero, it follows that the energy due to the velocity is all imparted to the

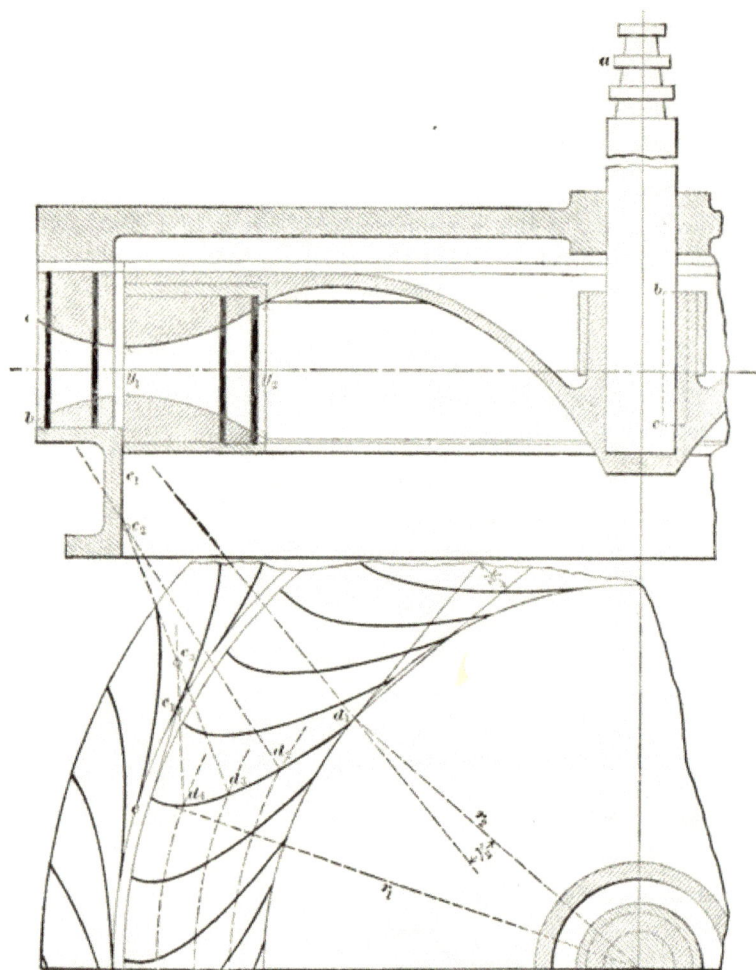

RADIALLY INFLOW TURBINE.

wheel; and the energy due to the remaining half of the head is imparted to the wheel by pressure in the wheel. If the velocity of entrance to the wheel be that due to the head, or $V^2 = 2gH$ then will no energy be imparted to the wheel on account of pressure exerted by any part of the head H, but if $V^2 < 2gH$, then will some of the work be done by this pressure, w being zero. For the cases in Table I., *the energy imparted to the wheel will be due one-half to velocity and one-half to pressure;* or in symbols,

$$U = \tfrac{1}{2}MV^2 + W \cdot \tfrac{1}{2}H$$

$$= \tfrac{1}{2}\frac{W}{g} \cdot gH + \tfrac{1}{2}WH = WH, \quad . \quad . \quad (28)$$

or, the entire potential energy of the water will be expended in work upon the wheel.

Whenever $V^2 < 2gH$, the pressure at entrance must exceed the external pressure at exit, and

$$\frac{H - \dfrac{V^2}{2g}}{H}, \quad . \quad . \quad . \quad . \quad . \quad . \quad (29)$$

then will be the part of the head producing pressure in the wheel.

In practice, a cannot be zero and is made from 20° to 30°. When other elements of the wheel are fixed, the value of a may be determined so as to secure a certain amount of initial pressure in the wheel, as will be shown hereafter.

The value $r_1 = 1.4r_2$ makes the width of the crown for internal flow about the same as for $r_1 = \sqrt{\tfrac{1}{2}}r_2$ for outward flow, being approximately 0.3 of the external radius.

9. *Values of* μ_1 *and* μ_2.

The frictional resistances depend not only upon the construction of the wheel as to smoothness of the surfaces, sharpness of the angles, regularity of the curved parts, but also upon the manner it is run; for if run too fast, the initial elements of the wheel will cut across the stream of water, producing eddies and preventing the buckets from being filled, and if run too slow, eddies and whirls may be produced and thus the effective sections be reduced. These values cannot be definitely assigned beforehand, but Weisbach gives for good conditions,

$$\mu_1 = \mu_2 = 0.05 \text{ to } 0.10. \quad \ldots \quad (30)$$

They are not necessarily equal, and μ_1 may be from 0.05 to 0.075, and μ_2 from 0.06 to 0.10, or values near these.

10. *Values of* γ_1.

It has already been shown that γ_1 must be less than $180^\circ - \alpha$. If $\gamma_1 = 90^\circ$, equation (14) shows that the efficiency of the frictionless wheel will be independent of α. The effect of different values for γ_1 is best observed from numerical results as shown in the following table:

TABLE II.

INITIAL ANGLE. γ_1 (1)	Let $\alpha = 25^\circ$,	$\gamma_2 = 12^\circ$,	$\mu_1 = \mu_2 = 0.10$.			
	$r_1 = r_2\sqrt{\frac{1}{2}}$			$r_1 = 1.4 r_2$.		
	$\omega'r_2$. (2)	E. (3)	$\omega'r_1$. (4)	$\omega'r_2$. (5)	E. (6)	$\omega'r_1$. (7)
60°	$1.322\sqrt{gH}$	.812	$.934\sqrt{gH}$	$.780\sqrt{gH}$	.911	$1.092\sqrt{gH}$
90°	1.226 "	.827	$.866$ "	$.689$ "	.908	$.964$ "
120°	1.078 "	.838	$.762$ "	$.576$ "	.898	$.806$ "
150°	$.518$ "	.744	$.366$ "	$.271$ "	.752	$.379$ "

The values $\omega'r_2$ in columns (2) and (5) are velocities for the terminal rim, which in column (2) are for the exterior rim, but

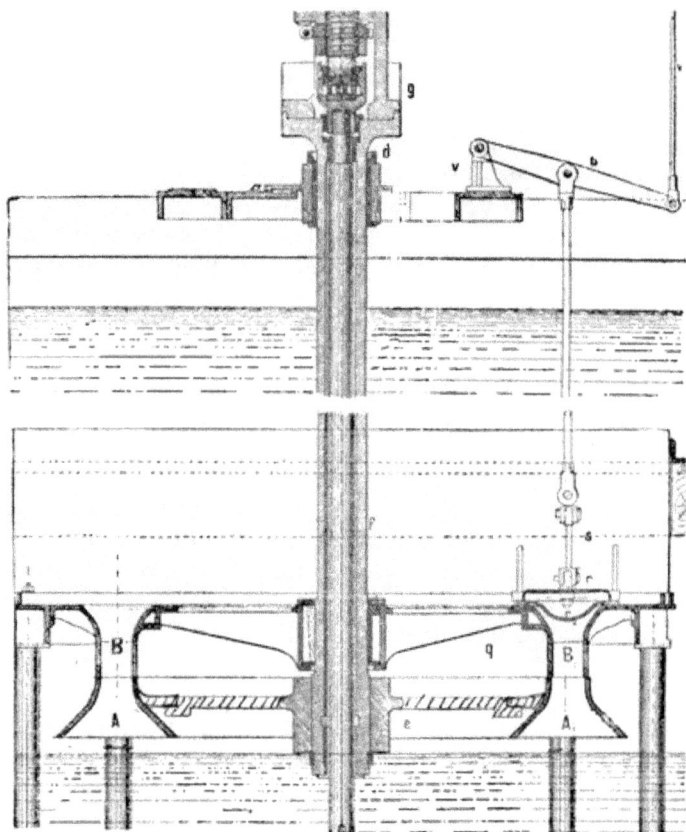

PARALLEL FLOW TURBINE, GIRARD TYPE.

for column (5) it is the interior rim, while column (7) is for the exterior rim.

Columns (2) and (7) show that the velocity of the outer rim is less, for maximum effect, for the inflow than for the outflow, for the same size wheel.

Column (3) shows that the efficiency, E, decreases as the initial angle of the bucket, γ_1, increases up to 120°. This maximum will be for this wheel with this amount of friction.

Column (6) shows that for the inflow wheel the efficiency continually decreases as γ_1 increases. If the head and quantity of water discharged be constant, the work would be proportional to the efficiency; for, from equation (14),

$$U = \delta Q H E \quad . \quad . \quad . \quad . \quad . \quad . \quad (31)$$

The effect of γ_1 on the velocities is shown in Table III.

TABLE III.

Let $\alpha = 25°$, $\qquad \gamma_2 = 12°$, $\qquad \mu_1 = \mu_2 = 0.10$, $\qquad Q = 1$.

INITIAL ANGLE. γ_1	$r_1 = r_2\sqrt{\tfrac{1}{4}}$						$r_1 = 1.4\, r_2$					
	$\dfrac{V}{\sqrt{gH}}$	$\dfrac{v_1}{\sqrt{gH}}$	$\dfrac{v_2}{\sqrt{gH}}$	$\dfrac{K\times}{\sqrt{gH}}$	$\dfrac{k_1\times}{\sqrt{gH}}$	$\dfrac{k_2\times}{\sqrt{gH}}$	$\dfrac{V}{\sqrt{gH}}$	$\dfrac{v_1}{\sqrt{gH}}$	$\dfrac{v_2}{\sqrt{gH}}$	$\dfrac{K\times}{\sqrt{gH}}$	$\dfrac{k_1\times}{\sqrt{gH}}$	$\dfrac{k_2\times}{\sqrt{gH}}$
60°	.820	.396	1.447	1.219	2.525	.691	.959	.463	.761	1.043	2.160	1.314
90°	.955	.403	1.378	1.047	2.481	.725	1.063	.449	.676	.940	2.227	1.479
120°	1.150	.560	1.153	.869	1.785	.874	1.217	.593	.605	.821	1.686	1.653
150°	2.100	1.775	.621	.476	.563	1.610	2.060	1.741	.296	.485	.574	3.378

For commercial considerations it may be necessary to sacrifice some efficiency to save on first cost, and to avoid making the wheel unwieldy.

From equation (4) it appears that the pressure in the wheel at entrance, p_1, diminishes as the velocity of admission, V, increases, and, according to equation (5), V depends upon γ_1 when α is fixed. Since the crowns are not fitted air tight nor water tight it is desirable that p_1 should exceed the pressure of the atmosphere when the wheel runs in free air, or the pressure $p_2 \div p_a$ when submerged, to prevent air or water from flowing in at the edge of the crown. It will be shown hereafter, in discussing the pressures in the wheel, that we should have

$$- \tan \gamma_1 > \tan 2\alpha, \quad \ldots \quad \ldots \quad \ldots \quad (32)$$

or, $$180° - \gamma_1 > 2\alpha,$$
or, $$\gamma_1 < 180° - 2\alpha.$$
If $$\alpha = 30,$$
then $$\gamma_1 < 120°.$$

To be on the safe side, the angle γ_1 may be 20 or 30 degrees less than this limit, giving

$$\gamma_1 = 180° - 2\alpha - 25 \text{ (say)}$$
$$= 155 - 2\alpha.$$

Then if $\alpha = 30°$, $\gamma_1 = 95°$. Some designers make this angle 90°, others more, and still others less than that amount. Weisbach suggests that it be less so that the bucket will be shorter and friction less. This reasoning appears to be correct for the inflow wheel, for the size and conditions shown in Table II., but not for the outflow wheel. In the Tremont turbines, described in the *Lowell Hydraulic Experiments*, this angle is 90°, the angle α, 20°, and γ_2, 10°. Fourneyron made $\gamma_1 = 90°$, and α from 30° to 33°.

In Table III. it appears that for $\gamma_1 = 150°$, $V = 2.1 \sqrt{gH}$, which exceeds $\sqrt{2gH}$; that is, the velocity of exit from the supply chamber exceeds that due to the head, hence the pressure at entrance into the wheel must be less than that of the atmosphere. For zero pressure for the frictionless wheel, the above condition gives

$$\gamma_1 = 180° - 2\alpha,$$

PARALLEL FLOW. JONVAL WITH DRAFT TUBE.

which for $\alpha = 25°$, gives $\gamma_1 = 130°$, and for $\nu_1 = 150°$, the pressure would be negative, and for $120°$ it would be positive. It appears that for the wheel *with friction*, considered in the table, that this pressure is also positive for $\gamma_1 = 120°$, and negative for $150°$.

11. *Form of Bucket.*

The form of the bucket does not enter the analysis, and therefore its proper form cannot be determined analytically. Only the initial and terminal directions enter directly, and from these and the volume of the water flowing through the wheel, the area of the normal sections may be found from equations (20).

But well-known physical facts determine that the changes of curvature and section must be gradual, and the general form regular, so that eddies and whirls shall not be formed. For the same reason the wheel must be run with the correct velocity to secure the best effect; for otherwise the effective angles α and γ_1 may be changed to values which cannot be determined beforehand, in which case the wheel cannot be correctly ana-

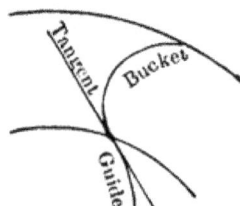

FIG. 5.

FIG. 6.

lyzed. In practice the buckets are made of two or three arcs of circles mutually tangential at their points of meeting. Also, if the normal sections, K, k_1, k_2, of the buckets as constructed do not agree with those given by computation, the stream will, if possible, adjust itself to true conditions by the formation of

2

eddies. If the terminal sections at the guides, or the initial
section of the bucket, be too small, the action may be changed
from a *pressure wheel* to one of *free deviation.* So long as the
pressure in the wheel exceeds the external pressure, the preceding analysis is applicable for the wheel running for best
effect, observing that the sections K, k_1, k_2, are not those of
the wheel, but those which are. computed from the velocities
V, r_1, v_2.

12. *Value of* θ; or direction of the quitting water.
From Fig. 1 it may be found that

$$V_2 \cos \theta = v_2 \cos \gamma_2 - \omega' r_2, \quad \ldots \ldots \quad (33)$$

and $\qquad V_2 \sin \theta = v_2 \sin \gamma_2; \quad \ldots \ldots \ldots \quad (34)$

$$\therefore \cot \theta = \cot \gamma_2 - \frac{\omega' r_2}{v_2 \sin \gamma_2}. \quad \ldots \ldots \quad (35)$$

These formulas are for the velocity giving maximum efficiency. If the speed be assumed, ω in place of ω' becoming
known, v_2 is given by equation (19). It is apparent for such a
case that θ may have a large range of values from $\theta = \gamma_2$, when
the wheel is at rest, to θ exceeding 90° for high velocities. The
following table gives some results :

TABLE IV.

γ_1	$\alpha = 25°$,	$\gamma_2 = 12°$,	$\mu_1 = \mu_2 = 0.10$.	
	$r_1 = r_2 \sqrt{4}$.		$r_1 = 1.4\, r_2$	
	ω	θ	ω	θ
60°	.314 $\sqrt{gH}$	72° 14'	.160 $\sqrt{gH}$	102° 43'
90°	.310 "	66° 59'	.143 "	101° 17'
120°	.241 "	60° 24'	.126 "	82° 52'
150°	.157 "	55° 26'	.043 "	74° 51'

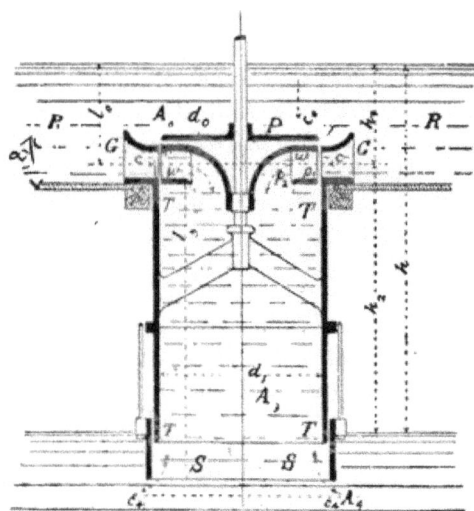

INFLOW PARALLEL CROWNS.

According to this table the water is thrown backward, or in the direction opposite to the motion of the wheel for the outward flow wheel, and for the inflow it is thrown forward for γ_1 less than 90°, and backward for γ_1 greater than 120°.

In the *Tremont turbine* a device was used for determining the direction of the water leaving the wheel, and for the best efficiency, 79¼ per cent., the angle θ was about 120°. *Lowell Hydraulic Experiments*, p. 33.

The angle thus observed had a large range of values ranging from 50° to 140° for efficiencies only two or three per cent. less than 79¼ per cent.

13. *Of the value of ω.*

So far as analysis indicates, the wheel may run at any speed; but in order that the stream shall flow smoothly from the supply chamber into the bucket—thus practically maintaining the angles α and γ_1—the relations in equations (5) and (6) must be maintained, or

$$v_1 = \frac{\sin \alpha}{\sin \gamma_1} V, \quad \ldots \quad \ldots \quad (36)$$

and this requires that the velocity V shall be properly regulated, which can be done by regulating the head h_1 or the pressure p_1 or both h_1 and p_1, as shown by equation (4). This however is not practical. In practice, the speed is regulated, and when the condition for maximum efficiency is established, the velocities V and v_1 are found from equations (17) and (18).

Since γ_2, in practice, is small we have, for best effect,

$$v_2 = \omega' r_2, \text{ approximately}, \quad \ldots \quad \ldots \quad (37)$$

and, adopting this value, a more simple expression may be found for the velocity of the wheel. For equation (19) gives

$$v_2 = r_2 \omega'$$
$$= \frac{\sqrt{gH}}{\sqrt{\dfrac{\cos \alpha \sin \gamma_1}{\sin(\alpha+\gamma_1)}\left(\dfrac{r_1}{r_2}\right)^2 + \frac{1}{2}\mu_1 \dfrac{\sin^2 \gamma_1}{\sin^2(\alpha+\gamma_1)}\left(\dfrac{r_1}{r_2}\right)^2 + \frac{1}{2}\mu_2}} . \text{(Approx.) } (38)$$

If $\mu_1 = \mu_2 = 0.10$, $r_2 \div r_1 = 1.40$, $\alpha = 25°$, $\gamma_2 = 90°$, the velocity of the *initial* rim for outward flow will be

$$\omega' r_1 = \frac{\sqrt{gH}}{\sqrt{1 + 0.159}} = 0.929 \sqrt{gH}.$$

The velocity due to the head would be

$$v_h = \sqrt{2gH} = 1.414 \sqrt{gH};$$

hence, the velocity of the initial rim should be about

$$\frac{0.928 \sqrt{gH}}{1.414 \sqrt{gH}} = 0.659 \quad . \qquad (39)$$

of the velocity due to the head.

For an inflow wheel in which $r_1^2 = 2r_2^2$, and the other dimensions, as given above, this becomes

$$\frac{0.954}{1.414} = 0.689 \quad . \quad . \quad . \quad . \quad . \quad (40)$$

of the velocity due to the head.

The highest efficiency of the Tremont turbine, found experimentally, was 0.79375, and the corresponding velocity, 0.62645 of the velocity due to the head, and for all velocities above and below this value the efficiency was less. Experiment showed that the velocity might be considerably larger or smaller than this amount without diminishing the efficiency very much.

In the Tremont turbine it was found that if the velocity of the initial (or interior) rim was not less than 44 nor more than 75 per cent. of that due to the fall, the efficiency was 75 per cent. or more. *Exp.*, p. 44.

This wheel was allowed to run freely without any brake except its own friction, and the velocity of the initial rim was observed to be $1.335 \sqrt{2gH}$, half of which is

$$0.6675 \sqrt{2gH}, \quad . \quad . \quad . \quad . \quad . \quad (41)$$

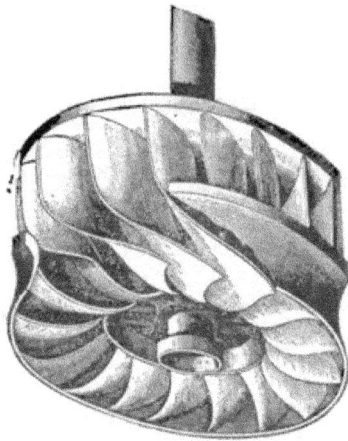

THE "RISDON."

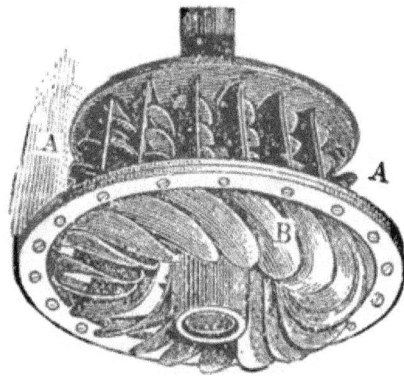

THE "HERCULES."

MIXED FLOW, INWARD AND DOWNWARD.

" which is not far from the velocity giving maximum effect; that is to say, *when the gate is fully raised the coefficient of effect is a maximum when the wheel is moving with about half its maximum velocity.*" *Exp.*, p. 37.

M. Poncelet computed the theoretical useful effect of a certain turbine of which M. Morin had determined the value by experiment. The following are the results (*Comptes Rendus*, 1838, *Juillet*):

TABLE V.

Velocity of initial rim or $\dfrac{r_1\omega'}{\sqrt{2gH}} =$	Number of turns of the wheel per minute.	Ratio of useful to theoretical effect.	Means of values by experiment.
0.0	0 00	0.000	. . . .
0.2	33.80	0.664	. . . .
0.4	47.87	0.773	0.700
0.6	58.61	0.807	0.705
0.7	62.81	0.810	0.700
0.8	67.67	0.806	0.675
1.0	75.76	0.786	0.610
1.2	82.88	0.753	0.490
1.4	89.52	0.712	0.360
1.6	95.70	0.664	0.280
1.8	101.51	0.612	0.203
2.0	107.00	0.546	0.050
3.72	145.00	0.000	. . . .

Poncelet states that he took no account of passive resistances, and hence his results should be larger than those of experiment as they are ; but here both theory and experiment give the maximum efficiency for a velocity of about 0.6 that due to the head, and the efficiency is but little less for velocities perceptibly greater and less than that for the best effect. For velocities considerably greater and less, theoretical results are much larger than those found by experiment, for reasons already given, chief of which is the fact that eddies are induced, and the effective angles of the mechanism changed to unknown values.

14. *Pressure in the wheel.*

Dropping the subscript $_2$ from v, r, p, in equation (9), the resulting value of p will give the pressure per unit at any point of the bucket providing that μ_2 be considered constant. Changing r to ρ, equation (9) thus gives

$$p = \left[v_1^2 - (1 + \mu_2) v^2 + \alpha^2 (\rho^2 - r_1^2) \right] \frac{\delta}{2g} + p_1. \quad . \quad (42)$$

To solve this requires a knowledge of the transverse sections of the stream, for the velocity v will be inversely as the cross section.

From equations (20) and (6)

$$v = v_1 \frac{k_1}{k} = \frac{k_1}{k} \cdot \frac{\sin \alpha}{\sin (\alpha + \gamma_1)} \cdot \omega r_1 \quad . \quad . \quad . \quad (43)$$

From (4) and (5),

$$p_1 = \delta h_1 + p_a - \delta \frac{1 + \mu_1}{2g} \cdot \frac{\sin^2 \gamma_1}{\sin^2 (\alpha + \gamma_1)} \, \omega^2 r_1^2. \quad . \quad (44)$$

These reduce equation (42) to

$$\frac{\mathcal{P}}{\delta} = \frac{\mathcal{P}_a}{\delta} + gh_1 + \tfrac{1}{2} \, \omega^2 \rho^2 - \tfrac{1}{2} \, \omega^2 r_1^2$$

$$- \frac{1}{2} \left[(1 + \mu_1) \sin^2 \gamma_1 + [k_1^2(1 + \mu_2) - k^2] \frac{\sin^2 \alpha}{k^2} \right] \frac{\omega^2 r_1^2}{\sin^2 (\alpha + \gamma_1)}. (45)$$

The back or concave side of the bucket will be subjected to a pressure which may be considered in two parts: one due to the deflection of the stream passing through it, the other to a pressure which is the same as that against the crown, and is uniform throughout the cross section of the bucket, due to the pressure of a part (or all) of the head in the supply chamber. It is the latter pressure which is given by the value of p in equation (45). The construction of the wheel being known, the pressure p may be found at any point of the wheel for any assumed practical velocity; although, for reasons previously

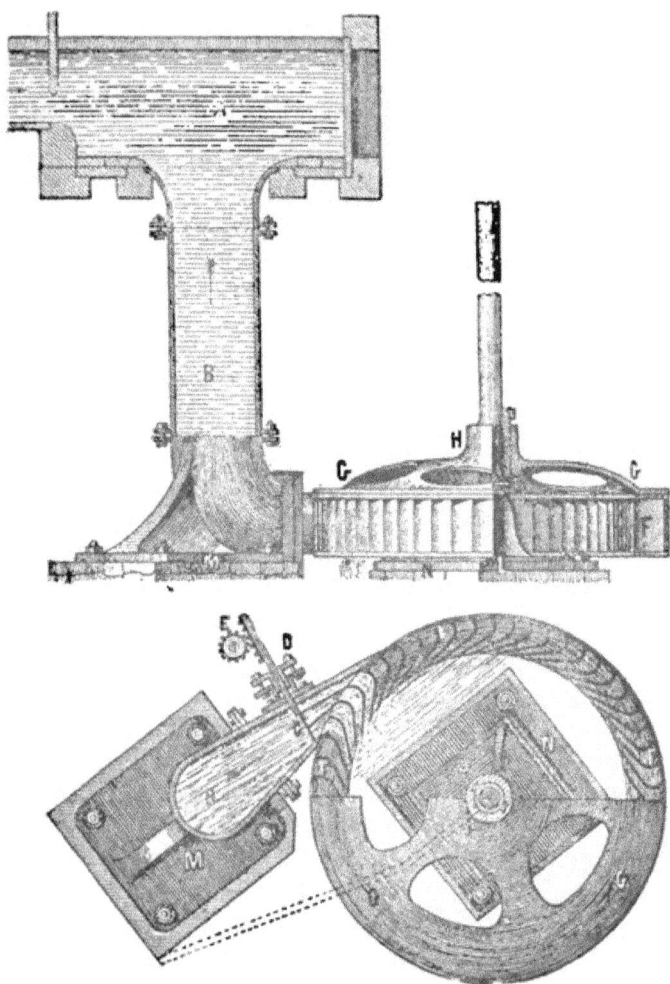

SEGMENTAL FEED.   RADIALLY INWARD FLOW.   TANGENTIAL WHEEL.

given, it will be of practical value only when running near the velocity for maximum efficiency. " There are two cases :

1. That in which the discharge is into free air ;
2. That in which the wheel is submerged.   •

In the first case if the pressure is uniform, the case is called that of 'free deviation' in which the entire pressure upon the forward side of the bucket is due to the deviation of the water from a right line, and will be considered further on.

If equation (45) shows a continually decreasing pressure from the initial element to that of exit, or if the minimum pressure exceeds p_a, the preceding analysis is applicable. But if it shows a point of minimum pressure less than p_a, it will be in a condition of unstable equilibrium, in which the slightest inequality would cause air to rush in and restore the pressure to that of the atmosphere; so that the pressure in the wheel and the flow would be changed. The point of minimum pressure may be found by plotting results found from equation (45), substituting values for ρ taken from measurements of the wheel, and k from computation. From the entrance of the wheel up to the point of minimum pressure the preceding analysis applies ; and the remainder of the wheel must be analyzed for 'free deviation' and the two results added.

In the second case the equations will apply, since air cannot enter, provided that p does not become negative, to realize which requires a tensile stress of the water. This is impossible and eddies would be formed ; and the effect of these on the velocity and pressure cannot be computed. Such a case cannot be analyzed."

15. *To find the pressure at the entrance to the bucket* when running at best effect. In (45) let $\rho = r_1$, $k = k_1$ and $p = p_1$. To simplify still more, let the wheel be frictionless, or $\mu_1 = \mu_2 = 0$, and find from equation (38)

$$r_1^2 \omega'^2 = \frac{\sin{(\alpha + \gamma_1)}}{\cos{\alpha} \sin{\gamma_1}} gH, \quad . \quad . \quad . \quad (46)$$

also $h_1 = H + h_2$, and (45) becomes

$$p_1 = \delta H + \delta h_2 + p_a - \frac{\delta \sin \gamma_1}{2 \cos \alpha \sin (\alpha + \gamma_1)} H. \qquad (47)$$

If the wheel is not submerged $h_2 = 0$, and let the pressure p_1 equal that of the atmosphere, or p_a, then

$$0 = 1 - \frac{\sin \gamma_1}{2 \cos \alpha \sin (\alpha + \gamma_1)}. \qquad \cdot \qquad \cdot \qquad (48)$$

If the wheel be submerged, let $p_1 = \delta h_2 + p_a$, and the equation reduces to that of the preceding.

Equation (48) gives

$$\tan 2\alpha = - \tan \gamma_1,$$

or, $$2\alpha = 180° - \gamma_1; \qquad \cdot \qquad \cdot \qquad (49)$$

for which value the pressure at the entrance to the wheel will equal that just outside.

If, $$2\alpha > 180° - \gamma_1, \qquad \cdot \qquad (50)$$

the pressure within will be less than that without ; but if

$$2\alpha < 180° - \gamma_1, \qquad \cdot \qquad (51)$$

the pressure within will exceed that without—a condition which is considered desirable. If frictional resistances be considered the value of $r_1 \omega'$ from equation (38) will be less than that given by (46), and hence the last term of equation (47) will be less unless α be greater than the value given by equation (51); hence with frictional resistances the terminal angle of the guide blade may exceed somewhat $90° - \frac{1}{2}\gamma_1$; therefore, if the value of α be found for a frictionless wheel it will be a safe value when there is friction. If $\gamma = 90°$ and $\alpha = 90° - \frac{1}{2}\gamma_1 = 45°$, then (47) gives

$$p_1 = \delta H + \delta h_2 + p_a - \delta H = \delta h_2 + p_a, \qquad \cdot \qquad \cdot \qquad (52)$$

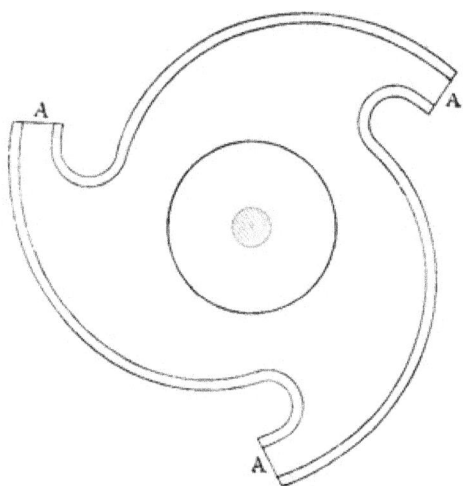

SCOTTISH OR WHITELAW.

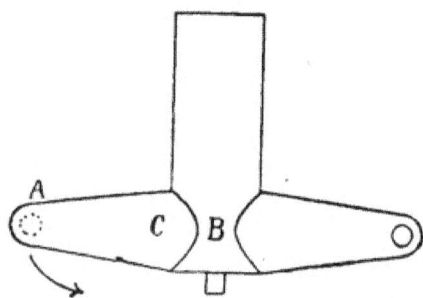

BARKER MILL.

WHEELS WITHOUT GUIDES.

as it should. If $\gamma_1 = 90°$ and $\alpha = 30°$, then

$$p_1 = \delta H + \delta h_a + p_a - \tfrac{3}{8}\delta H,$$

or,

$$p_1 > \delta h_2 + p_t + 0.336 H. \qquad \qquad . \quad . \quad . \quad . \quad (53)$$

The angle α should not be so small or γ_1 so small as to produce excessive pressure at the entrance to the wheel.

Example.—Find the pressure per square inch at the entrance to the wheel when the head is 10 feet, the terminal angle of the guide is 30°, the initial angle of the bucket $\gamma_1 = 90°$; the wheel being one foot under the water in the tail race.

16. *Number of buckets.*

The analysis given above is true for a wheel with a single bucket, provided the supply is constantly open to the bucket and closed by the remainder of the wheel. But for practical considerations the wheel should be full of buckets, although the number cannot be determined by analysis. Successful wheels have been made in which the distance between the buckets was as small as 0.75 of an inch, and others as much as 2.75 inches. *Lowell Hyd. Exp.*, p. 47. Turbines at the Centennial Exposition had buckets from 4½ inches to 9 inches from centre to centre.

17. *Ratio of radii.*

Theory does not limit the dimensions of the wheel. In practice,

for outward flow, $r_2 \div r_1$ is from 1.25 to 1.50 ⎞
for inward flow, $r_2 \div r_1$ is from 0.66 to 0.80 ⎠ $\cdot \cdot$ (54)

It appears from Table II. that the inflow wheel has a higher efficiency than the outward flow wheel (columns 6 and 3), and these wheels have about the same outside and inside diameters. The inflow wheel also runs somewhat slower for best effect. The centrifugal force in the outward flow wheel tends to force the water outward faster than it would otherwise flow; while in the inward flow wheel it has the contrary effect, acting as it does in opposition to the velocity in the buckets.

It also appears that the efficiency of the outward flow wheel increases slightly as the width of the crown is less, and the velocity for maximum efficiency is slower; while for the inflow wheel the efficiency slightly increases for increased width of crown and the velocity of the outer rim at the same time also increases.

$$\text{Let } r_1 = n r_2. \; \gamma_1 = 90^\circ, \; \gamma_2 = 20^\circ, \; \mu_1 = \mu_2 = 0, \; \alpha = 30^\circ ;$$
$$\text{then for } n = 0, \qquad 0.5, \qquad 0.8, \qquad 1.4,$$
$$\text{we have } \omega' r_1 = 0, \qquad 0.761 g H, \; 0.972 g H, \; 1.005 g H,$$
$$\omega' r^2 = 1.391 g H, \; 1.522 g H, \; 1.215 g H, \; 0.718 g H,$$
$$E = 0.6594, \qquad 0.8050, \qquad 0.9070, \qquad 0.9784.$$

18. *Efficiency*, E.

The method of determining the theoretical value of E has already been given; but to determine the actual value, resort must be had to experiments. These have been made in large numbers and the results published. By assuming the minimum values of the several losses, a maximum limit to the efficiency may be fixed. Thus, if the actual velocity be 0.97 of the theoretical, the energy lost will be $(1 - 0.97^2)$ or 6 per cent.

Friction along the buckets and bends . . . 5 " "
Energy lost by impact, say . . . . 2 " "
Energy lost in the escaping water . . . . 3 " "

                       Total . . . . 16 " "
                  Leaving . . . . . 84 " "

available for work. This discards the friction of the mechanism and frictional losses along the guides, and if 2 per cent. be allowed for the latter, there will be left 82 per cent. It

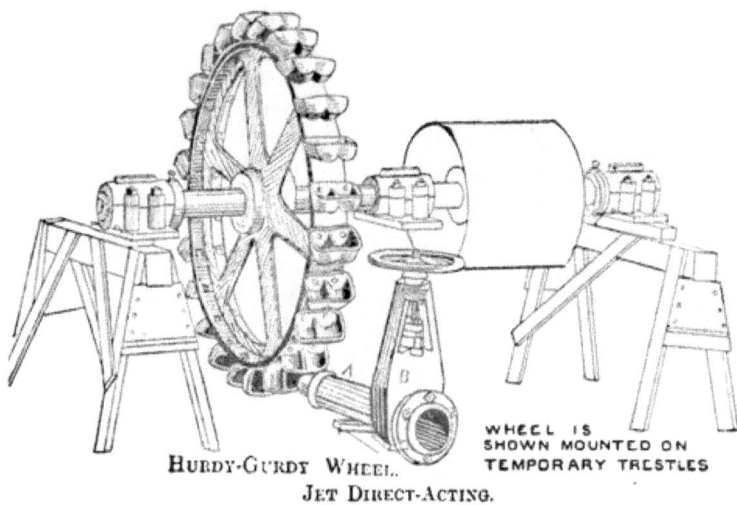

HURDY-GURDY WHEEL.
JET DIRECT-ACTING.

WHEEL IS
SHOWN MOUNTED ON
TEMPORARY TRESTLES

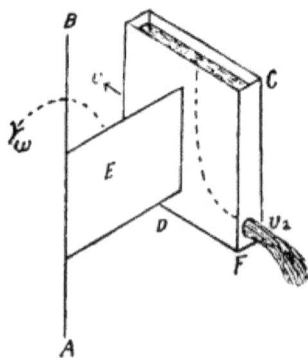

JET REACTION

JET WHEELS.

seems hardly possible for the effective efficiency to exceed 82
per cent., and all claims of 90 or more per cent. for these
motors should be at once discarded as being too improbable
for serious consideration. A turbine yielding from 75 to 80
per cent. is extremely good. The celebrated Tremont turbine
gave 79 per cent. *Lowell Exp.*, p. 33. Experiments with
higher efficiencies have been reported. A Jonval turbine
(parallel flow) was reported as yielding 0.75 to 0.90, but Morin
suggested corrections reducing it to 0.63 to 0.71. (Weisbach,
Mech. of Eng., vol. ii., p. 501.) Weisbach gives the results of
many experiments, in which the efficiency ranged from 50 to
84 per cent. See pages 470, 500–507. See also *Jour. Frank.
Inst.*, 1843, for efficiencies from 64 to 75 per cent. Numerous
experiments give $E = 0.60$ to 0.65. The efficiency, considering
only the energy imparted to the wheel, will exceed by several
per cent. the efficiency of the wheel, for the latter will include
the friction of the support, and leakage at the joint between
the sluice and wheel, which are not included in the former; also
as a plant the resistances and losses in the supply chamber are
to be still further deducted.

19. *The Crowns.*—The crowns may be plane annular discs,
or conical, or curved. If the partitions forming the buckets be
so thin that they may be discarded, the law of radial flow will
be determined by the form of the crowns. If the crowns be
plane, the radial flow (or radial component) will diminish as
the distance from the axis increases—the buckets being full—
for the annular space will be greater.

20. *Designing.*
The dimensions of a wheel must be determined for a definite
velocity. Thus far it has been assumed that the angles α, γ_1,
etc., are given, and the normal sections of the stream thus
deduced. We will now assume that all the dimensions of
the buckets are known, and the angle α and the section K are

to be determined. The velocities v_1 and v_2 must now be found independently of a. From Fig. 1 we have

$$V^2 = v_1^2 + \omega^2 r_1^2 - 2v_1\omega r_1 \cos \gamma_1 \qquad \cdot \qquad (55)$$

which combined with equations (4), (9), (10), $v_1 k_1 = v_2 k_2$, as in (20), and $H = h_1 - h_2$, will give

$$v_2 = \frac{k_1}{k_2} v_1 = \frac{(1 + \mu_1) k_1 k_2 \omega r_1 \cos \gamma_1}{(1 + \mu_2) k_1^2 + \mu_1 k_2^2}$$

$$+ \sqrt{\frac{2gH + \left[r_2^2 - (2 + \mu_1) r_1^2 \right] \omega^2}{(1 + \mu_2) k_1^2 + \mu_1 k_2^2} k_1^2 + \left[\frac{(1 + \mu_1) k_1 k_2 \omega r_1 \cos \gamma_1}{(1 + \mu_2) k_1^2 + \mu_1 k_2^2} \right]^2}$$

$$= A\omega + \sqrt{2gHB + C^2\omega^2} \qquad \cdot \qquad \cdot \qquad (56)$$

where

$$A = \frac{(1 + \mu_1) k_1 k_2 r_1 \cos \gamma_1}{(1 + \mu_2) k_1^2 + \mu_1 k_2^2}, \qquad \cdot \qquad \cdot \qquad \cdot \qquad (56a)$$

$$B = \frac{k_1^2}{(1 + \mu_2) k_1^2 + \mu_1 k_2^2} \qquad \cdot \qquad \cdot \qquad \cdot \qquad (56b)$$

$$J^2 = \frac{r_2^2 - (2 + \mu_1) r_1^2}{(1 + \mu_2) k_1^2 + \mu_1 k_2^2} k_1^2 \qquad \cdot \qquad \cdot \qquad (56c)$$

$$C^2 = J^2 + A^2 \qquad \qquad \cdot \qquad (56d)$$

$$v_1 = \frac{k_2}{k_1} v_2 = cv_2 \qquad \cdot \qquad \cdot \qquad \cdot \qquad \cdot \qquad (57)$$

Equations (11), (12), (13), (55), (56), and (57), after making

$$\begin{rcases} a = 1 + \mu_2 + \mu_1 c^2 \\ b = r_2 \cos \gamma_2 + \mu_1 c r_1 \cos \gamma_1 \end{rcases}, \qquad \cdot \qquad (57a)$$

give

$$E = \frac{U}{\delta Q H} = \frac{1}{gH} \left[gH (1 - aB) - \left[a \left(A^2 + \tfrac{1}{2} J^2 \right) - bA + \tfrac{1}{2} r_2^2 + \tfrac{1}{2} \mu_1 r_1^2 \right] \omega^2 + (b - aA) \, \omega\sqrt{2gHB + C^2\omega^2} \right] \cdot (58)$$

Remark.

Since μ_1 is a comparatively small fraction and k_1 exceeds k_2 certain terms may be omitted giving for v_2 the approximate but very nearly exact value,

$$v_2 = \frac{1 + \mu_1}{1 + \mu_2} \cdot \frac{k_2}{k_1} \, \omega r_1 \cos \gamma_1$$

$$+ \sqrt{\frac{2gH + r_2^2 - (2 + \mu_1)\, r_1^2}{1 + \mu_2} \, \omega^2 + \left[\frac{1 + \mu_1}{1 + \mu_2} \cdot \frac{k_2}{k_1} \, \omega r_1 \cos \gamma_1 \right]^2}$$

and corresponding reductions in A, B, J.

Let

$$D = gH (1 - aB),$$
$$F^2 = a (A^2 + \tfrac{1}{2} J^2) - bA + \tfrac{1}{2} r_2^2 + \tfrac{1}{2} \mu_1 r_1^2,$$
$$G = b - aA.$$

Then

$$E = \frac{1}{gH} \left[D - F^2 \omega^2 + G\omega \sqrt{2gHB + C^2\omega^2} \right] \quad \cdot \quad (59)$$

For a maximum $dE \div d\omega = 0$, giving

$$\omega^2 = gHB \frac{F^2 - \sqrt{F^4 - C^2 G^2}}{C^2 \sqrt{F^4 - C^2 G^2}} \quad \cdot \quad \cdot \quad \cdot \quad (60)$$

This value of ω is the one to be used in the other equations. The form of equation (60) is the same as that of equation (16). Substituting in (59), observing that $aB = 1$, and hence $D = 0$, we find

$$E_{max} = \left(F^2 - \sqrt{F^4 - C^2 G^2} \right) \frac{B}{C^2} \quad \cdot \quad \cdot \quad (61)$$

$$\therefore \omega^2 = \frac{gH E_{max}}{\sqrt{F^4 - C^2 G^2}} \quad \cdot \quad \cdot \quad \cdot \quad \cdot \quad (60a)$$

To find the terminable angle α of the guide blade that will enable the stream to flow smoothly, subject to the preceding conditions, Fig. 1 gives

$$V \cos \alpha = \omega r_1 - v_1 \cos \gamma_1,$$

which, combined with equation (55), gives

$$\cos \alpha = \frac{\omega r_1 - v_1 \cos \gamma_1}{\sqrt{V_1^2 + \omega^2 r_1^2 - 2v_1 \omega r_1 \cos \gamma_1}}. \quad \cdot \quad \cdot \quad (62)$$

Eliminating v_1 by means of equations (57) and (56) gives $\cos \alpha$ in terms of the six constants r_1, r_2, γ_1, γ_2, k_1, and k_2, which are fixed and known from the dimensions of the wheel, and of the velocity ω of the wheel. Since the wheel may run at different velocities the angle α must vary, and this will be done in practice by the piling of the water in the passages. Each turbine, however, should be designed to run at the speed giving maximum efficiency, and its angles and dimensions should satisfy equations (60) and (62).

From equation (9),

$$2g\,\frac{p_1}{\delta} = (1 + \mu_2)\,v_2^2 - v_1^2 - \omega^2\,(r_2^2 - r_1^2) + 2g\,\frac{p_a + \delta h_2}{\delta},\quad . \quad (63)$$

in which, if v_2 and v_1 be substituted from above, p_1 becomes known.

Similarly, V from equation (55) becomes known, and finally, from (60),

$$\cos \alpha = \frac{\omega r_1 - v_1 \cos \gamma_1}{V}. \quad . \quad . \quad . \quad . \quad (64)$$

21. *Path of the Water.*—Let aA be the position of the bucket when the water enters at b. The bucket being drawn in position to a scale, divide it into any number of parts—equal or unequal—aa_1, a_1a_2, etc., and find the time required for it to go from a to a_1. The distance being small, assume that the velocity is uniform from a to a_1, and equal to v_1, which will be given by equation (20),

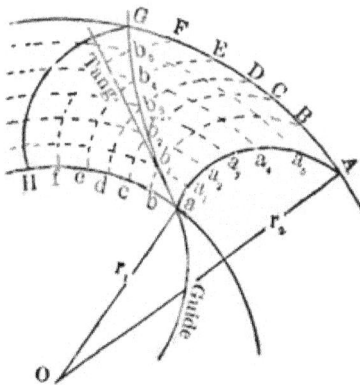

$$v = \frac{Q}{k}.$$

or better,

$$v = \frac{k_1}{k}\,v_1 \quad . \quad . \quad (64a)$$

FIG. 7.

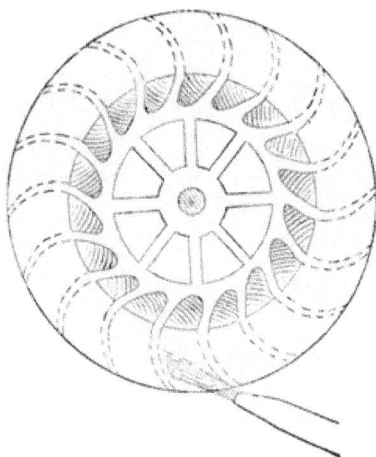

THE
COLLINS
WHEEL

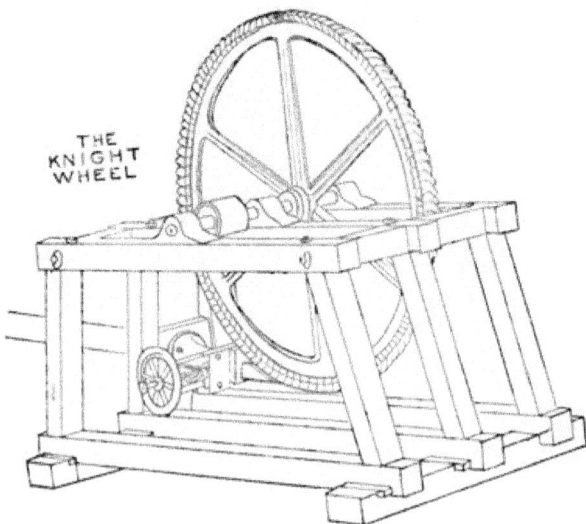

THE
KNIGHT
WHEEL

Then will the time t be

$$t = \frac{a a_1}{v} \qquad \ldots \ldots \ldots (65)$$

During this time the rim has gone from a to b a distance

$$ab = r_1 \, \omega t. \qquad \ldots \ldots \ldots (66)$$

If the bucket bB be drawn through b, and the arc $a_1 b_1$ through a_1, their intersection b_1 will be the position of the particle at the end of the time t. In a similar manner, the successive points b_2, b_3, etc., may be found, through which a continuous curve may be drawn representing the path of the stream.

The line tangent to the termination of the bucket, will indicate the direction of the water at entrance of the wheel, and if the water drives the wheel, the *path* should be entirely outside this line and convex toward it.

22. *Design a guide blade, outflow turbine.*

Assume the effective fall, ... $H =$ ft.
Assume the required horse-power, . . . $HP =$
Assume the exit angle, ... $y_2 =$
Assume the entrance angle of bucket, . . $y_1 = 90°$.
Fix the exit angle of the guide, Eqs. (32), (51), $a = 30°$.?
Assume efficiency, ... $E = 0.65$, or **0.70**;
and after the wheel is fully designed, re-
  compute this value and if necessary
  correct the dimensions.
Required quantity of water per second
  without loss, ... $Q = U \div \delta H$.
Required quantity, ... $Q \div E =$
Assume $\mu_1 = 0.10$, $\mu_2 = 0.075$.
Velocity of the initial rim, Eq. (40), approx., $\omega r_1 =$
  (The corrected, final value will be found
  by Eq. (16) or (60).

Let $r_2 = 1.3r_1$, then velocity of outer rim, $\omega r_2 =$
The velocity into the bucket, Eq. (18), . . $v_2 =$
Initial section of buckets, Eq. (20), . . . $k_1 =$
The inner circumference will be $2\pi r_1$. Let
the walls of the buckets be $\frac{1}{16}$ of the
circumference, then the effective open-
ings will be $\frac{13}{16}$ of the circumference, or
$\frac{13}{8}\pi r_1$. Assume a depth, y, between
the crowns. Try $y = \frac{1}{2}r_1$. Then will
the initial cross section of all the
buckets be $\frac{13}{16}\pi r_1^2$; hence, $\frac{13}{16}\pi r_1^2 = k_1$; $\therefore$ $r_1 =$
If the radius is not what is desired, it may
be changed to some other value and
the depth y computed. Then, . . . $r_2 =$
The cross section at outer rim will be, if
the crowns are planes, $\frac{13}{8}\pi r_2\, y \sin \gamma_2 = k_2 =$
The number of buckets will be assumed . $=$

Having determined these elements, the final velocity v_2 in
reference to the bucket may be computed by equation (67),
ωr_1 from (62), V from (55), and α from (64).

If the turbine revolves in air, at least half the depth of the
wheel is to be deducted from the head H.

If the circular opening between the wheel and gate be $\frac{1}{16}$
of an inch, or $\frac{1}{192}$ of a foot, and the coefficient of discharge be
0.7, the discharge will be

$$\frac{1}{192}\, 0.7 \times 2\pi r_1 \times \sqrt{\frac{2gp_1}{\delta}} = q, \quad . \quad . \quad . \quad (67)$$

p_1 being determined from equation (63). The loss of work
will be

$$62.2q \times H \quad \text{or} \quad 62.2q\,(H - \tfrac{1}{2}y). \quad . \quad . \quad . \quad (68)$$

The work lost by friction, if the radius of the axle be r_3, the

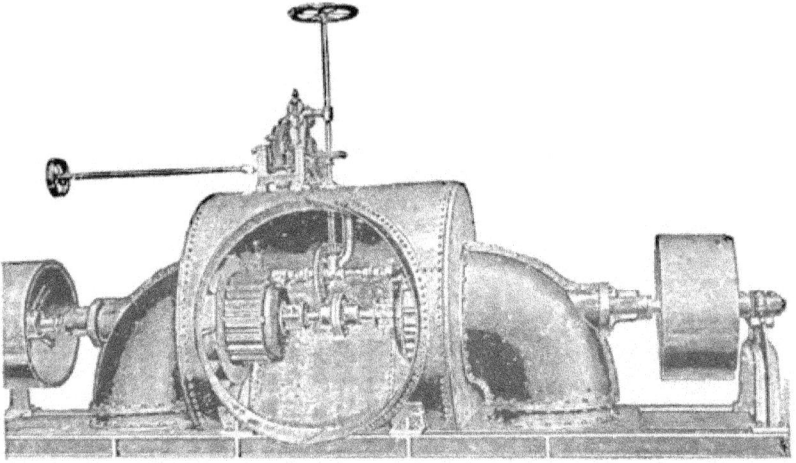

TWIN TURBINES ON A HORIZONTAL AXIS.

weight of the loaded wheel, W_3, and coefficient of friction μ_3, will be nearly

$$\frac{2}{3}\,\mu_3\,W_3\,r_3\omega \text{ per sec.} \quad . \quad . \quad . \quad . \quad . \quad (69)$$

The work done by the water must be the effective U plus the work due to the losses. The work done by the water passing through the wheel must be U plus that given by equation (69). Call this U_1. Compute the work done by the water, and let it be U_2; then will the required depth be

$$y_2 = \frac{U_1}{U_2}y \quad . \quad . \quad . \quad . \quad . \quad . \quad (70)$$

The efficiency may now be recomputed.

The following tables, VI. and VII., though for wheels of special dimensions, give some general results as shown in the following "Conclusions." The sections k_1 and k_2 are assumed to be those of the wheel, and further it is assumed that the wheel passages (or channels) are filled with flowing water, and hence without eddies.

TABLE VI.

OUTWARD FLOW TURBINE. $\quad r_1 = r_2 \sqrt{\tfrac{1}{2}} \quad \mu_1 = \mu_2 = 0.10. \quad \gamma_2 = 12°. \quad$ Parallel crowns. $\quad k_1 r_1 = k_2 r_2 = KV = Q = 1.$

Initial angle. γ_1	Efficiency E_r Eq. (61).	Velocity outer rim. $r_1\omega'$ Eq. (60).	Velocity inner rim. $r_1\omega' = \frac{1}{2}\sqrt{\tfrac{1}{2}} \, r_2\omega'$	Relative velocity of exit. v_2 Eq. (56).	Relative velocity of entrance. v_1 Eq. (57).	Velocity of exit from supply chamber. V_1 Eq. (55).	Terminal angle of guide. α Eq. (62).	Direction of quitting water. θ Eq. (83).	Head equivalent of energy in quitting water. $\frac{V_2^2}{2g}$	$k_1 r_1$ Eqs. (30), (56).	$A_1 \sqrt{g H}$ Eqs. (30), (57).	$K \sqrt{g H}$ Eqs. (30), (55).
1	2	3	4	5	6	7	8	9	10	11	12	13
60°	0.801	$1.354 \sqrt{g H}$ $0.952 \sqrt{2g H}$	$0.952 \sqrt{g H}$ $0.895 \sqrt{2g H}$	$1.483 \sqrt{g H}$ $1.018 \sqrt{2g H}$	$0.508 \sqrt{g H}$ $0.386 \sqrt{2g H}$	$0.842 \sqrt{g H}$ $0.595 \sqrt{2g H}$	$34° 17'$	$78°$	$0.051 H$	0.67	1.984	1.187
90°	0.828	$1.236 \sqrt{g H}$ $0.874 \sqrt{2g H}$	$0.874 \sqrt{g H}$ $0.619 \sqrt{2g H}$	$1.317 \sqrt{g H}$ $0.931 \sqrt{2g H}$	$0.385 \sqrt{g H}$ $0.574 \sqrt{2g H}$	$0.956 \sqrt{g H}$ $0.654 \sqrt{2g H}$	$25° 56'$	$79°$	$0.039 H$	0.73	2.584	1.046
120°	0.841	$1.129 \sqrt{g H}$ $0.796 \sqrt{2g H}$	$0.798 \sqrt{g H}$ $0.563 \sqrt{2g H}$	$1.192 \sqrt{g H}$ $0.843 \sqrt{2g H}$	$0.605 \sqrt{g H}$ $0.967 \sqrt{2g H}$	$1.000 \sqrt{g H}$ $0.819 \sqrt{2g H}$	$19° 5'$	$82°$	$0.031 H$	0.81	2.444	0.943
150°	0.921	$1.005 \sqrt{g H}$ $0.709 \sqrt{2g H}$	$0.709 \sqrt{g H}$ $0.501 \sqrt{2g H}$	$0.999 \sqrt{g H}$ $0.707 \sqrt{2g H}$	$0.587 \sqrt{g H}$ $0.416 \sqrt{2g H}$	$1.252 \sqrt{g H}$ $0.886 \sqrt{2g H}$	$19° 31'$	$95°$	$0.022 H$	1.00	1.703	0.799

TABLE VII.

INWARD FLOW TURBINE. $\quad r_1 = \sqrt{2}\, r_2 \quad \mu_1 = \mu_2 = 0.10. \quad \gamma_2 = 12°. \quad$ Parallel crowns. $\quad k_1 r_1 = k_2 r_2 = KV = Q = 1.$

γ_1	E_r Eq. (61).	Velocity outer rim. $r_1\omega'$ Eq. (60).	Velocity inner rim. $r_2\omega'$	v_2 Eq. (56).	v_1 Eq. (57).	V Eq. (55).	α Eq. (62).	θ Eq. (83).	$\frac{v^2}{2g}$	$k_2 r_2$ Eqs. (30), (56).	$A_1 \sqrt{g H}$ Eqs. (30), (57).	$K \sqrt{g H}$ Eqs. (30), (55).
1	2	3	4	5	6	7	8	9	10	11	12	13
60°	0.920	$1.003 \sqrt{g H}$ $0.709 \sqrt{2g H}$	$0.709 \sqrt{g H}$ $0.501 \sqrt{2g H}$	$0.673 \sqrt{g H}$ $0.476 \sqrt{2g H}$	$0.115 \sqrt{g H}$ $0.489 \sqrt{2g H}$	$0.951 \sqrt{g H}$ $0.672 \sqrt{2g H}$	$7° 0'$	$110°$	$0.010 H$	1.28	8.85	1.05
90°	0.920	$0.973 \sqrt{g H}$ $0.688 \sqrt{2g H}$	$0.688 \sqrt{g H}$ $0.487 \sqrt{2g H}$	$0.664 \sqrt{g H}$ $0.470 \sqrt{2g H}$	$0.084 \sqrt{g H}$ $0.069 \sqrt{2g H}$	$0.957 \sqrt{g H}$ $0.691 \sqrt{2g H}$	$5° 28'$	$106°$	$0.010 H$	1.50	10.39	1.02
120°	0.919	$0.915 \sqrt{g H}$ $0.698 \sqrt{2g H}$	$0.698 \sqrt{g H}$ $0.473 \sqrt{2g H}$	$0.645 \sqrt{g H}$ $0.456 \sqrt{2g H}$	$0.109 \sqrt{g H}$ $0.077 \sqrt{2g H}$	$1.001 \sqrt{g H}$ $1.000 \sqrt{2g H}$	$4° 46'$	$105°$	$1.000 H$	1.55	9.17	0.99
150°	0.918	$0.896 \sqrt{g H}$ $0.634 \sqrt{2g H}$	$0.634 \sqrt{g H}$ $0.448 \sqrt{2g H}$	$0.607 \sqrt{g H}$ $0.429 \sqrt{2g H}$	$0.178 \sqrt{g H}$ $0.125 \sqrt{2g H}$	$1.053 \sqrt{g H}$ $0.743 \sqrt{2g H}$	$9° 08'$	$107°$	$0.009 H$	1.65	5.95	0.95

Conclusions.—From an examination of Tables II., III., VI., VII., the following conclusions are drawn:

1. The maximum theoretical efficiency of the inflow wheel is perceptibly larger than that of the outflow, the width of crowns and the initial and terminal angles of the buckets being the same. One reason for this is due to the flow through the wheel being opposed by the centrifugal action, but more particularly to the smaller velocity of discharge from the inflow wheel.

2. Columns (10) in Tables VI. and VII. show that for the wheels here considered the loss of energy due to the quitting velocity is from 2.2, 5.1 per cent. from the outflow, and from 0.9 to 1 per cent for the inflow.

3. The same tables show that in column (2) the efficiency is almost constant for the varying conditions here considered, while for the outflow there is considerable variation.

4. One of the most interesting and profitable studies to the theorist and practitioner is the effect upon the efficiency due to properly proportioning the terminal angle, α, of the guide blade. It will be observed that all the efficiencies in Tables VI. and VII. exceed the corresponding ones in Table II. except the first in column (3) of Table II. In Table II. the terminal angle, α, is constantly 25°, while in Tables VI. and VII. it is less than that value, and in the highest efficiencies very much less.

5. It appears from these tables that the terminal angle, α, has frequently been made too large for best efficiency.

6. That the terminal angle, α, of the guide should be comparatively small for best effect: for the inflow less than 10°, and that theoretically, when the angle is about 7°, the efficiency is some 10 per cent. greater than when it is 25° in the wheels here considered.

7. Tables II. and VI. indicate that the initial angle of the bucket should exceed 90° for best effect for outflow wheels.

8. Tables II. and VII. show that the initial angle should be less than 90° for best effect on inflow wheels, but that from 60° to 120° the efficiency varies scarcely 1 per cent.

9. The most marked effect in properly proportioning the terminal angle, α, of the guide is shown when the initial angle of the bucket is 150°. In this case the efficiency for the outflow when α is 25° is 0.744, Table II., but when α is 13½°, as in Table VI., it becomes 0.921. For the inflow, in the former case, it is 0.752, but when the angle is 3°, as in Table VII., it becomes 0.918.

10. Since the wheels here considered have the same width of crowns and the same terminal angle of the bucket, the depths of the wheels will be proportional to k_2 for discharging equal volumes of water. Tables III., VI., VII. show that the section k_2 increases as the initial angle of the buckets increases, and that it must be greater for the inflow than for the outflow; hence the depth of the wheel must be greater for the inflow for delivering the same volume of water.

11. But the same volume of water delivered by the inflow does more work than that of the outflow; the depths should be as k_2 divided by the efficiency. Thus in Tables VI. and VII., for $\gamma = 90°$, and for the same heads, H, the relative depths should be for equal works $(0.759 \div 0.828) \div (150 \div 0.920) = 1.67$.

12. In the outflow wheel, column (9), Table VI., shows that for the outflow for best effect the direction of the quitting water in reference to the earth should be nearly radial (from 76° to 97°), but for the inflow wheel the water is thrown forward in quitting (column [9] Table VII.). This alone shows that the velocity of the rim should somewhat exceed the relative final velocity backward in the bucket, as shown in columns (4) and (5).

13. In these tables I have given all the velocities in terms of $\sqrt{2\,g\,h}$, and the coefficients of this expression will be the part of the head which would produce that velocity if the water issued freely. In Tables VI. and VII. there is only one case, column (5) of the former table, where the coefficient exceeds unity, and the excess is so small it may be discarded; and it may be said that in a properly proportioned turbine with the conditions here

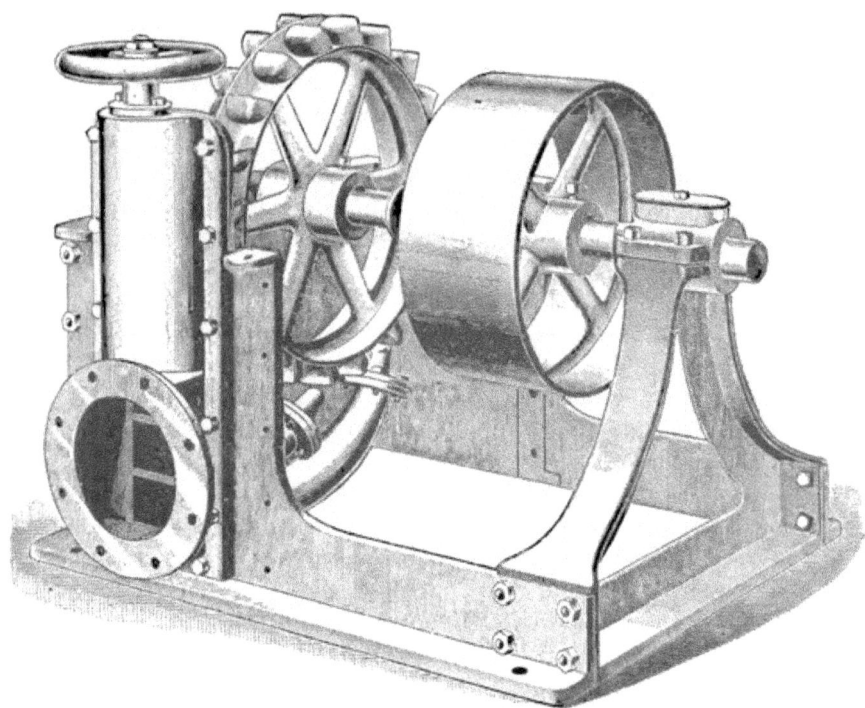

CASCADE WHEEL.

given, none of the velocities will equal that due to the head in the supply chamber when running at best effect.

14. The inflow turbine presents the best conditions for construction for producing a given effect, the only apparent disadvantage being an increased first cost due to an increased depth, or an increased diameter for producing a given amount of work. The larger efficiency should, however, more than neutralize the increased first cost.

15. Column 7 and equation (29) show that the pressure at the initial rim decreases as the initial angle γ_1 increases.

16. Tables VI. and VII. are for parallel crowns. Examples of buckets of variable depths will be given later, and are illustrated in Figs. 19 and 24.

SPECIAL WHEELS.

23. *Fourneyron Turbine.*—All wheels having guide blades are of the Fourneyron type, although the wheels made by him were outward flow. The preceding analysis is a general solution of this turbine.

24. *Francis and Thomson's vortex wheels* are inward flow wheels with guide blades. The preceding analysis is also applicable to these wheels.

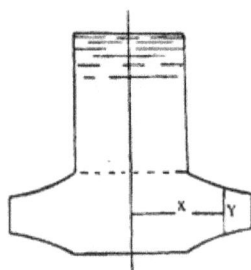

Fig. 8.

25. *The Jonval Turbine* is a parallel flow wheel with guide blades, to which the preceding analysis is applicable by making $r_1 = r_2$.

(For the details of these and many other forms, see *Hydraulic Motors* by Weisbach.)

26. *Rankine Wheel.*—This is a wheel of the Fourneyron type, but Rankine having made certain modifications in its

assumed construction it is indicated by his own name.
(Fig. 8.)

It is an outflow wheel and the crowns are so made that
the *radial* velocity of the water in passing through the wheel
will be uniform. If x be the abscissa from the axis of the
wheel to any point of the crown, and y the distance between
the crowns at that point, v_r the radial component of the
velocity, then

$$y \cdot 2\pi x \cdot v_r = Q,$$

or, $$yx = a \ constant, \ . \ . \ . \ . \ (71)$$

which is the equation to an hyperbola referred to its asymp-
totes. This determines the form of the crowns. If the wheel
were inward flow, the depth would be greatest at the inner
rim.

In this wheel the initial element of the bucket is radial, or
$\gamma_1 = 90°$; and Rankine *assumed* that the velocity for best effect
must be such that the water will quit the wheel radially, or
$\theta = 90°$. These conditions given, from Fig. 1 and equations
(5), (6), (34), for a *frictionless* wheel,

$$\omega r_1 = V \cos \alpha. \ . \ . \ . \ . \ . \ . \ (72)$$

$$\omega r_2 = v_2 \cos \gamma_2. \ . \ . \ . \ . \ . \ (73)$$

$$v_1 = w = V \sin \alpha = \omega r_1 \tan \alpha = \omega r_2 \tan \gamma_2 = v_2 \sin \gamma_2; \quad (74)$$

$$\therefore \ \tan \alpha = \frac{r_2}{r_1} \tan \gamma_2. \ . \ . \ . \ . \ (75)$$

which determines the proper angle of the guide blade when
the value of γ_2 has been assigned. If $\gamma_2 = 15°$, $r_2 = 1\frac{1}{2}r_1$, then
$\alpha = 18\frac{1}{2}°$; and if $\gamma_2 = 20$, $r_2 = 1.3$, then $\alpha = 24°$ nearly. But
to be certain that the internal pressure exceeds the external,
α should not exceed these values. Equations (19), (73), and (75)
give

$$\omega r_1 = \sqrt{\frac{2gH}{2 + \tan^2 \alpha}}, \ . \ . \ . \ . \ (76)$$

which establishes the velocity of the initial rim of the wheel.

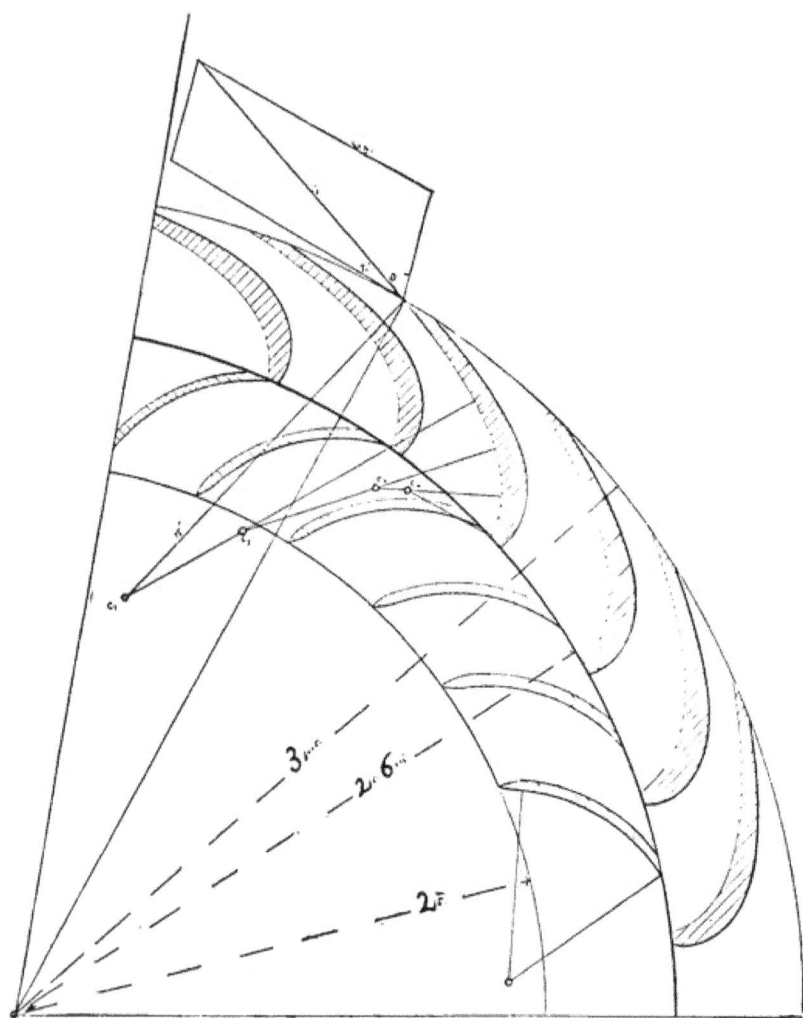

PLAN OF A PROPOSED FOURNEYRON WHEEL.

The work in the frictionless wheel will be the theoretical work the water could do, less the energy in the water quitting the wheel, or

$$U = WH - \tfrac{1}{2}\frac{W}{g}w^2$$

[Eq. (74)]
$$= WH - \tfrac{1}{2}\frac{W}{g}r_1^2\omega^2\tan^2\alpha. \quad \ldots \quad (77)$$

[Eq. (76)]
$$= \frac{WH}{1+\tfrac{1}{2}\tan^2\alpha}. \quad \cdots \cdots \quad (78)$$

The efficiency will be

$$E = \frac{U}{WH} = \frac{1}{1+\tfrac{1}{2}\tan^2\alpha}. \quad \cdots \quad (79)$$

27. The following is to show that Rankine's assumption of velocity for best efficiency is not quite correct. Substituting $\gamma_1 = 90^\circ$, $\mu_1 = 0$, $\mu_2 = 0$, $r_1 = nr_2$, in equation (16) gives

$$\omega^2 r_2^2 = \frac{gH}{1-2n^2}\left[\frac{1-n^2-\sqrt{(1-n^2)^2-\cos^2\gamma_2(1-2n^2)}}{\sqrt{(1-n^2)^2-\cos^2\gamma_2(1-2n^2)}}\right] \quad (80)$$

and these in equation (19) will give

$$v_2\cos\gamma_2 = \sqrt{gH\cos^2\gamma_2\left[\frac{1-n^2+\sqrt{(1-n^2)^2-\cos^2\gamma_2(1-2n^2)}}{\sqrt{(1-n^2)^2-\cos^2\gamma_2(1-2n^2)}}\right]}$$
$$= \omega r_2\left[1-n^2+\sqrt{(1-n^2)^2-\cos^2\gamma_2(1-2n^2)}\right]. \quad (81)$$

This does not give
$$v_2\cos\gamma_2 = \omega r_2$$

as in equation (73), except when $n^2 = \tfrac{1}{2}$ (or $2r_1^2 = r_2^2$); and hence the direction of the water at exit will not be radial, as Rankine assumed, except for this case; and hence the analysis in article 26 is not applicable to the inflow, but for such wheels as have the proportion $2r_1^2 = r_2^2$, it is sufficiently exact for the friction-

less outflow wheel; and, as seen above, the hypothesis greatly
simplifies the analysis.

The condition for best efficiency of the frictionless wheel
requires that the velocity of leaving the wheel should be a
minimum; and this may be realized, in some cases, when its
direction is oblique to the radius.

Thus, let AC be radial when AB is the velocity relative to
the bucket, and BC the velocity of the rim; then it may be, in
some cases, that when AD is the relative velocity of exit, AE,
the velocity of exit relative to the earth, will be less than AC,
as shown in Fig. 9.

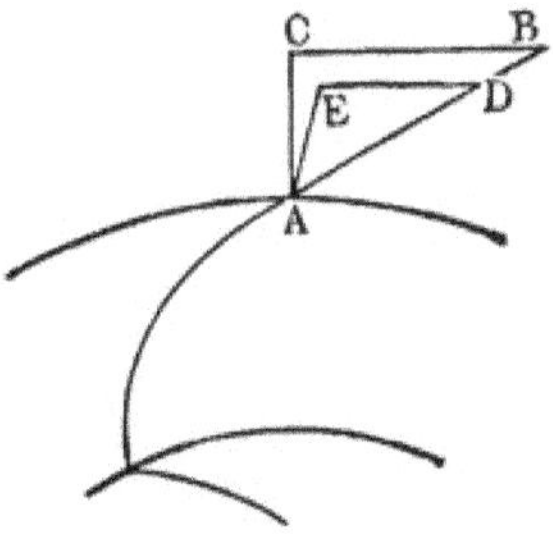

FIG. 9. 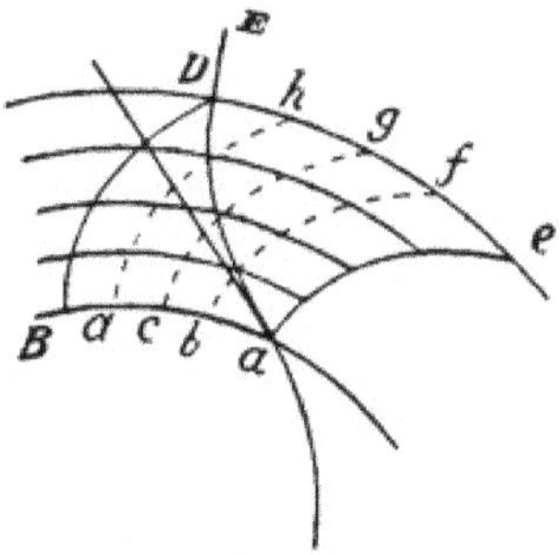   FIG. 10.

28. *The path of the water* is easily constructed for this wheel.
Since the radial velocity is uniform, the time of flowing through
the wheel will be

$$t = \frac{r_2 - r_1}{r_1}, \quad \cdots \cdots \quad (82)$$

during which time the initial rim AB, Fig. 10, will have travelled

$$aB = \omega r_1 t \text{ feet.} \quad \cdots \cdots \quad (83)$$

Divide $r_2 - r_1$ into equal parts by concentric arcs, and the
space aB into the same number of equal parts, and through
the points of division a, b, c, d, trace the buckets; then will aD,
drawn through the proper intersections of the arcs and buckets,
be the required path.

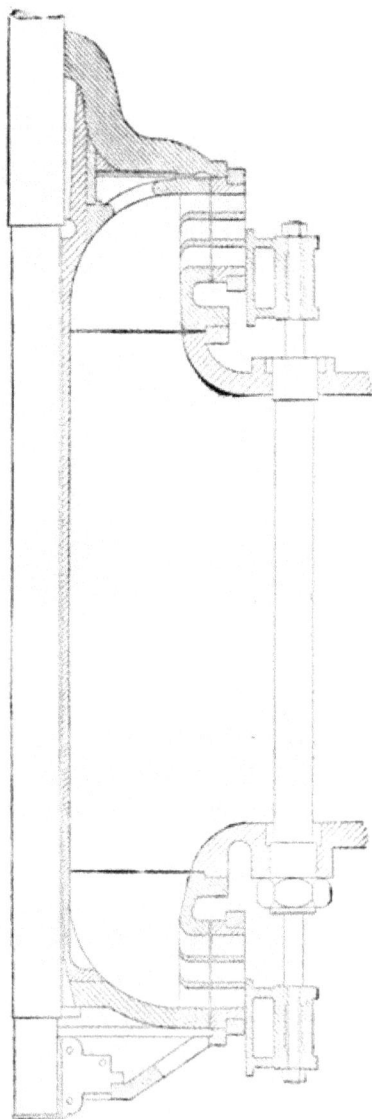

VERTICAL SECTION OF A PROPOSED DOUBLE FOURNEYRON WHEEL WITH TRIPLE CHAMBERS FOR GREAT FALLS. TRIPLE DOUBLE WHEEL.

9. Analyze a Rankine turbine, having given : $H = 12$ feet, $\gamma_2 = 15°$, $r_1 = 2$ feet, $r^2_2 = 2r^2_1$. Depth of outer rim, 6 inches.

Find Radius of outer rim, ... $r_2 =$

Angular velocity, ... $\omega =$

Velocity of initial rim, ... $r_1\omega =$

Velocity of outer rim, ... $r_2\omega =$

Angle of guide plates, ... $\alpha =$

Velocity from the supply chamber, . $V =$

Initial velocity in bucket, ... $v_1 =$

Terminal velocity in bucket, ... $v_2 =$

Velocity of quitting water, ... $w =$

Depth of inner rim, ... $y_1 =$

The horse-power, ... $HP =$

The efficiency, ... $E =$

If the partitions for the buckets occupy $\frac{1}{20}$ of the wheel, and the losses due to frictional resistances in the wheel and friction of the wheel be 20 per cent., what will be

The horse-power, ... $HP =$

The efficiency, ... $E =$

Find the pressure at the inner rim, . . . $p_1 =$

Find the path of the water.

29. *Velocity of a particle along a tube rotating about an axis perpendicular to its plane.*

This problem has already been solved in establishing the general equations of turbines, and the following is given to present it from another point of view.

If the particle at A, whose mass is m, be confined while the tube rotates about O, Fig. 11, with the angular velocity ω, the centrifugal force would be

$$f = m\omega^2\rho. \qquad . . . . \quad (84)$$

If θ be the angle between the radius vector prolonged and the normal upon the tangent, the component in the direction of the tangent to the tube will be

$$m\omega^2\rho \sin \theta,$$

and when the particle is free to move, this component will be effective for producing motion, and if the pressures at the opposite ends of the element are not equal, but differ by an amount dp, we have the equation

$$m\frac{d^2s}{dt^2} = m\omega^2\rho \sin \theta - dp \quad . \quad . \quad . \quad . \quad (85)$$

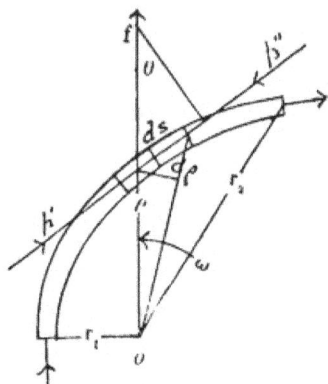

Fig. 11.

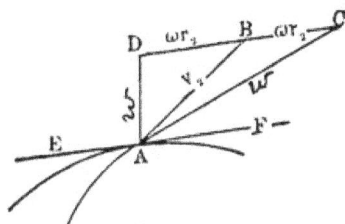

Fig. 12.

But $$ds \sin \theta = d\rho,$$

which combined with the preceding equation gives

$$\frac{ds\,d^2s}{dt^2} = \omega^2\rho d\rho - \frac{1}{m}\frac{d\rho}{\sin \theta}\,dp. \quad . \quad . \quad . \quad (86)$$

But $\dfrac{d\rho}{\sin \theta} = ds$, and if δ be the weight of unity of volume, then $m = \dfrac{\delta}{g}ds$, and the last term becomes $\dfrac{g}{\delta}dp$. The integral will be

$$\frac{ds^2}{dt^2}\bigg]_{v_1}^{v_2} = v_2^2 - v_1^2 = \omega^2\left(r_2^2 - r_1^2\right) - 2g\frac{p_2 - p_1}{\delta}. \quad . \quad (87)$$

If the friction be $\mu_2 v_2^2$, the equation becomes

$$(1 + \mu_2)\, v_2^2 = v_1^2 + \omega^2\,(r_2^2 - r_1^2) - 2g\,\frac{p_2 - p_1}{\delta}, \quad . \quad (88)$$

which is the same as equation (9).

If $p_2 = p_1$, and $\mu_2 = 0$, then

$$v_2^2 = v_1^2 + \omega^2\,(r_2^2 - r_1^2). \quad . \quad . \quad . \quad . \quad (89)$$

This gives the velocity relative to the tube whether it revolves to the right or left, and whatever be its curvature. If it revolves to the left, the resultant velocity will be AD, Fig. 12; if to the right, it will be AC. If γ_2 be measured from the arc backward of the motion, or $\gamma_2 = BAF$ for rotation to the left, and $\gamma_2 = EAB$ for motion to the right; then

$$AD^2 = w^2 = v_2^2 + \omega^2 r_2^2 - 2v_2 . \omega r_2 \cos \gamma_2. \quad . \quad . \quad (90)$$

$$AC^2 = w^2 = v_2^2 + \omega^2 r_2^2 + 2v_2 . \omega r_2 \cos \gamma_2. \quad . \quad . \quad (91)$$

In the latter case the quitting velocity will exceed the terminal velocity in the tube, and therefore increased velocity will have been imparted to the water—a condition requiring that energy be imparted to the wheel from an external source. In the former case the wheel is a motor, in the latter it is a receiver or transmitter of power; in the former the water drives the wheel, in the latter the wheel drives the water and virtually becomes a centrifugal pump.

If the water issues tangentially to the path described by the orifice, then $\gamma_2 = 0$, and

$$V_2 = v_2 \mp \omega r_2, \quad . \quad . \quad . \quad . \quad . \quad (92)$$

the upper sign belonging to the motor, and the lower to the pump.

Exercise.—If $r_1 = 1$ ft. $r_2 = 5$ ft. $\mu_2 = 0.1$, $v_1 = 5$ ft. per second, and the bucket rotates about a vertical axis 30 times per minute, and discharges the water directly backward, making $y_2 = 0$, required the terminal velocity along the tube and the velocity of discharge relatively to the earth.

30. *Wheel of Free Deviation.*—In this wheel the water in the buckets has a free surface, or, in other words, is subjected only to the pressure of the atmosphere. For this case

$$p_2 = p_1 = p_a; \ h_1 = H, \text{ and } h_2 = 0,$$

and equation (4) gives

$$(1 + \mu_1) \ V^2 = 2gH, \quad . \quad . \quad . \quad . \quad (93)$$

which will be the velocity of discharge from the supply chamber into the wheel; it is the velocity due to the head in the supply chamber when frictional resistance is included.

The triangle ABC, Fig. 1, gives

$$v_1^2 = V^2 + \omega^2 r_1^2 - 2 V \omega r_1 \cos \alpha, \quad . \quad . \quad . \quad (94)$$

which substituted in equation (88) gives

$$(1 + \mu_2) \ v_2^2 = V^2 + \omega^2 r_2^2 - 2 V . \omega r_1 \cos \alpha, \quad . \quad (95)$$

and this in equation (90) gives V_2, and equation (12) will give the required work. An exact general solution involves a solution of the general equation of the fourth degree. See article 89. The following is an approximate solution.

If y_2 be small, and the wheel be run for best effect, that is, so as to make the velocity V_2 very small, and considering $y_2 = 0$, equation (92) makes

$$v_2 = \omega r_2 \text{ nearly.}$$

Using this value as if it were the exact one, also neglecting friction, (95) gives

$$2V . \omega r_1 \cos \alpha = V^2 = 2gH,$$

or

$$2\omega r_1 \cos \alpha = \sqrt{2gH};$$

$$\therefore \omega r_1 = \frac{\sqrt{\tfrac{1}{2}gH}}{\cos \alpha} \cdot \text{(approx.)} \quad . \quad . \quad . \quad (96)$$

which gives the proper velocity of the initial rim ; and for the terminal rim

$$\omega r_2 = \frac{\sqrt{\frac{1}{2}gH}}{\cos \alpha} \cdot \frac{r_2}{r_1} \quad . \quad . \quad . \quad . \quad . \quad . \quad (97)$$

Number of revolutions per minute

$$N = 30 \frac{\omega}{\pi}. \quad . \quad . \quad . \quad . \quad . \quad . \quad (98)$$

To find the velocity at any point of the bucket relative to the bucket, drop the subscript $_2$ from equation (95) giving

$$(1 + \mu_2) v^2 = 2gH + \omega^2 r_2^2 - 2 \sqrt{2gH} . \omega r_1 \cos \alpha. \quad . \quad (99)$$

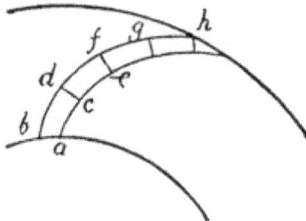

FIG. 13.

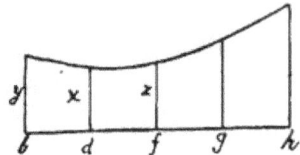

FIG. 14.

From **Fig. 1** or equations (4) and (5) find

$$\sin \gamma_1 = \frac{\sqrt{2gH}}{v_1} \sin \alpha. \quad . \quad . \quad . \quad . \quad (100)$$

In the frictionless wheel, the work done will be

$$U = \frac{1}{2} M (V^2 - V_2^2), \quad . \quad . \quad . \quad . \quad (101)$$

and the efficiency will be

$$E = \frac{U}{\delta QH} . \quad . \quad . \quad . \quad . \quad . \quad (102)$$

31. *To find the form of the free surface*, let the bucket be very narrow, so that a normal to one of the curves will be approximately normal to the other. Divide one side of the bucket

into any convenient number of parts, as ac, ce, etc., and erect
normals to the arc, as ab, cd, etc. Lay off these arcs on a right
line. Compute the velocity at any point, as d, Fig. 13, by formula (99). Let x be the required depth at d, then because the
velocity into the section equals q, the volume passing through
one of the buckets per second, we have

$$x.dc.v = q;$$

$$\therefore x = \frac{q}{dc.v}, \quad \ldots \quad \ldots \quad (103)$$

and similarly for all other sections. If only relative heights
are to be found, the quantity q need not be found, for if y be
the height at b, Fig. 14, then

$$y.ba.v_1 = q;$$

$$\therefore x = \frac{ba.v_1}{dc.v}y, \quad \ldots \quad \ldots \quad (104)$$

and by assuming any arbitrary value for y the relative value of
x becomes known. Similarly, the relative heights at all other
sections may be found.

32. *To find the path of the fluid* in reference to the earth, proceed as in Article 21 of the discussion of the general case.

33. *Exercise.*—Design a 30 horse-power inflow turbine of
free deviation, given an effective head of 16 feet.

Assume the depth of gate opening to be 4 inches ($\frac{1}{3}$ foot),
and after the computation has been completed if it does not
give 30 horse-power the depth may be changed by proportion.
Let the radius of the outer or initial rim be 1 ft.; of the inner
rim, $\frac{3}{4}$ of a foot; terminal angle of the bucket, $y_2 = 15°$; terminal angle of the guide, $\alpha = 30°$, $\mu_1 = 0.10 = \mu_2$.

Then, velocity of exit from supply chamber,

Eq. (93), ... $V =$
Velocity of outer rim, Eq. (96), ... $\omega r_1 =$
Velocity of inner rim, Eq. (97), ... $\omega r_2 =$

Number of turns per minute, ... $=$
Initial angle of bucket, Eq. (100), . . . $\gamma_1 =$
Initial velocity in bucket, Eq. (94), . . . $v_1 =$
Terminal velocity in bucket, Eq. (95), . . $v_2 =$
Velocity of exit, Eq. (90), ... $w =$
Direction of outflow, Eq. (35), ... $\theta =$
Coefficient of discharge 0.60, volume of
    water, ... $Q =$
Weight of water ($\delta = 62.4$), ... $\delta Q =$
Work per second, Eq. (101), ... $U =$
Horse-power, ... $HP =$
Efficiency, ... $E =$

If 90 per cent. of U is effective work, and if this does not give 30 horse-power, then the depth of the wheel should be

$$d = \frac{30}{0.9\,U}\ 4\ \text{inches.}$$

Find the profile of the stream in the buckets.

34. The following is taken from the report of the Commissioners of the Centennial Exposition, 1876, on Turbines, Group XX. The tests were for two minutes each. The revolutions and horse-powers here given are those corresponding to the best efficiencies :

Diameter of wheel. Inches.	Head in supply chamber. Feet.	Revolutions per minute.	Horse-power.	Efficiency, per cent.	No. of Buckets.	Kind of wheel.
30	31	255	95	85.0	10	Inflow.
24	31	302	67	77.0	14	Parallel.
24	30½	310	64	74.5	13	
27	30	291	76.8	80.3	16	
30	30	257	74	75.5	18	
25	31	288	46	82.0	12	Parallel.
30	29.2	258	80.5	78.7	13	In and down
25	30	279	62.5	83.7	15	In and down.
27	30.4	246	53.2	73.6	14	Parallel.
36	29.6.	197	66.2	83.8	26	Parallel.

These tests were by no means exhaustive. It is not known that they were run for best effect. The distance from centre to centre of buckets varied from 4.3 inches to 9.5, and at these extreme values the efficiencies were about the same. The number of gate openings was less than the number of buckets.

TURBINES WITHOUT GUIDES.

35. *Barker's Mill.*—As ordinarily constructed, this motor has two hollow arms connected with a central supply chamber,

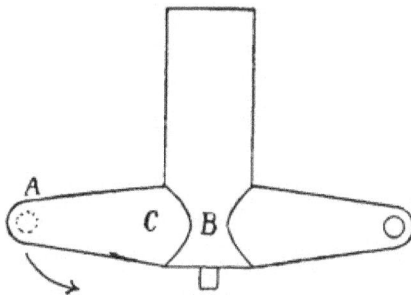

FIG. 15.

with orifices near their outer ends and on opposite sides of the arms. There are no guide plates The supply chamber rotates with the arms. The arms may be cylindrical, conical, or other convenient shape.

Since the water issues perpendicularly to the arms $\gamma_2 = 0$; and since the initial elements of the arms are radial, $\gamma_1 = 90°$, and as the water must flow radially into the arms, $\alpha = 90°$. The inner radius is necessarily small and may be considered zero. Hence, making

$$\gamma_2 = 0, \ \gamma_1 = 90°, \ \alpha = 90°, \ r_1 = 0,$$

equation (14) gives

$$E = \frac{U}{\delta Q H} = \frac{\omega r_2}{g H}\left[-\omega r_2 + \sqrt{\frac{2gH + \omega^2 r_2^2}{1 + \mu_2}}\right]. \quad . \quad (105)$$

Equation (19) gives

$$v_2 = \sqrt{\frac{2gH + \omega^2 r_2^2}{1 + \mu_2}}; \quad . \quad . \quad . \quad . \quad (106)$$

hence, the efficiency reduces to, for the frictionless wheel,

$$E = \frac{2\omega r_2}{v_2 + \omega r_2}. \quad \ldots \ldots \quad (107)$$

This has no algebraic maximum, but approaches unity as the velocity increases indefinitely. Practically it has been found that the best effect is produced when the velocity of the orifices is about that due to the head, or

$$\omega r_2 = \sqrt{2gH}; \quad \ldots \ldots \quad (108)$$

for which value the efficiency will be, if $\mu_2 = 0.10$

$$E = 2\left[-1 + \sqrt{\frac{2}{1.1}}\right] = 0.70. \quad \ldots \quad (109)$$

If k_2 be the area of the effective section of the orifice, then

$$Q = k_2 v_2. \quad \ldots \ldots \quad (110)$$

The pressure on the back side the arms opposite the orifices will be

$$P_1 = Mv_2 = \frac{\delta Q}{g} v_2. \quad \ldots \ldots \quad (111)$$

Of this pressure there will be required

$$P_2 = M.\omega r_2 = \frac{\delta Q}{g}.\omega r_2, \quad \ldots \ldots \quad (112)$$

to impart to the water the rotary velocity ωr_2 which it has when it reaches the orifice. The effective pressure will be $P_1 - P_2$, and the work done per second will be this pressure into the distance it traverses per second, or

$$U = \omega r_2 [P_1 - P_2],$$

which reduces to the value found from equations (105) and (106).

36. *Exercise.*—Let the supply chamber be square, and from two of its opposite sides let pyramidal arms project. Let

$H = 10$ feet, orifices each 2 square inches, vertical section of arms through the orifice each 4 square inches, section of the arms where they join the supply chamber each 3 square inches, horizontal section of the supply chamber 36 square inches, $r_2 = 36$ inches, velocity of the orifice $\omega r_2 = \sqrt{2gH}$, coefficient of discharge 0.64, and $\mu_2 = 0.10$.

Required :

Velocity of discharge relative to the orifice, $v_2 =$

Velocity of discharge relative to the earth, $V_2 =$

Velocity at entrance to the arms, . . . $v_1 =$

Velocity in the supply chamber, . . . $=$

The volume of water discharged, . . . $Q =$

The weight of water discharged, . . . $\delta Q =$

The work per second, ... $U =$

The horse-power, ... $HP =$

The efficiency, ... $E =$

The pressure on arm opposite orifice at A per square inch, ... $p_1 =$

The pressure at base of the arms at C, $p =$

The equation to the path of the fluid.

37. *Scottish and Whitelaw Turbines.*—These wheels have no guide plates, and differ from Barker's mill chiefly in having curved arms. The analysis is precisely the same as for the Barker mill. The only practical difference consists in providing a curved path for the water, instead of compelling the water to seek its path, forming eddies, etc.

38. *Jet Propeller.*—We first show how this problem may be solved by the preceding equations, and afterwards make an independent solution. Let a narrow vessel, Fig. 16, be carried by an arm E about a shaft $B.1$. Let water, by any suitable device, be dropped into the vessel, the horizontal velocity of the water being the same as that of the vessel. At F, the lower end of this chamber, let there be an orifice from which water may issue horizontally. The water may then be con-

sidered as entering the vessel or bucket without velocity, and passing downward finally curve towards, and issue from, the orifice. It thus becomes a parallel flow wheel without guides, and we have, for the frictionless wheel,

$$r_1 = r_2, \quad \gamma_1 = 90°, \quad \gamma_2 = 0, \quad \mu_1 = \mu_2 = 0, \quad H = 0,$$
$$p_2 = p_1 = p_a, \quad z_1 = 0$$

in equation (8); hence, the velocity of exit relative to the orifice will be

$$v_2^2 = 2gz_2 \quad \cdots \cdots \cdots \quad (113)$$

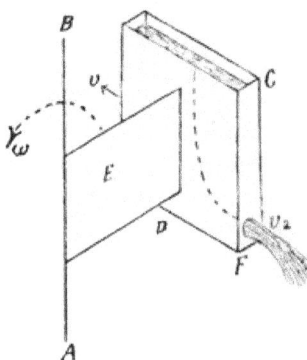

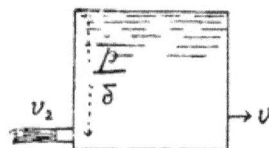

Fig. 17.

Fig. 16.

where z_2 equals the head in the supply chamber. Under these conditions the velocity of discharge will be independent of the velocity of rotation, if the rotation be uniform.

Equation (11) gives for the velocity of discharge relative to the earth

$$V_2 = v_2 - \omega r_2 \quad \cdots \cdots \cdots \quad (114)$$

Equations (19) and (14) give

$$U_1 = \frac{\delta Q}{g} v_2 \cdot r_2 \omega. \quad \cdots \cdots \cdots \quad (115)$$

This equation may be factored thus, $\dfrac{\delta Q}{g}$ is the mass of liquid flowing out per second; represent by M; Mv_2 is the momentum of the outflowing liquid per second. Mv_2r_2 is the moment of the momentum, and, finally,

$Mv_2 \cdot r_2 \omega$ is the moment of the momentum into the angular velocity, and equals the work done.

Let $v = \omega r_2 =$ the velocity of the vessel; then from (114) and (115)

$$w = v_2 - v, \quad \ldots \ldots \ldots \quad (116)$$

$$U_1 = Mv_2 v; \quad \ldots \ldots \ldots \quad (117)$$

which equations are true whether the motion be circular, linear, or in any other path.

In practice, the velocity of the jet is produced by the pressure exerted by a pump, in which case z_2 in equation (113) would be replaced by a virtual head, Fig. 17, equivalent to z_2; or

$$v^2_2 = 2g\frac{p}{\delta} \quad \ldots \ldots \ldots \quad (118)$$

Also the vessel, instead of having water supplied to it at the velocity of the vessel, picks it up from a body of water considered at rest; thus imparting to the water the momentum Mv, requiring the work per second

$$U_2 = Mv^2. \quad \ldots \ldots \ldots \quad (119)$$

Hence the effective work done by a jet propeller picking up the water from a state of rest will be

$$U = U_1 - U_2 = M(v_2 - v)v. \quad \ldots \ldots \quad (120)$$

The energy exerted by the pump will be that producing the velocity of water relative to the earth, or $\frac{1}{2}M(v_2 - v)^2$, plus that doing the work of driving the vessel; hence, the energy expended will be

$$\tfrac{1}{2} M(v_2 - v)^2 + U;$$

and the efficiency will be

$$E = \frac{U}{U + \tfrac{1}{2}M(v_2 - v)^2} = \frac{2v}{v_2 + v}. \quad . \quad . \quad . \quad (121)$$

This has no algebraic maximum, but approaches unity as v, the velocity of the vessel, in reference to the earth, approaches v_2 in value, the velocity of the jet in the opposite direction relative to the orifice.

If $v_2 = v$, *the efficiency will be perfect* as shown by (120), *but no work will be done* as shown by (119). This would be the case of a vessel drawn by an external agency, or even floating along a stream; for the water backward relative to the vessel would equal the forward velocity of the vessel.

The mass of the jet per second will vary as the section of the orifice and velocity of the jet; and if k be the section of the jet, then

$$U = \frac{\delta}{g} k v_2 (v_2 - v) v; \quad . \quad . \quad . \quad . \quad (122)$$

hence the same work may be done by enlarging the section k, and properly diminishing the velocity v_2 of the jet; but as v_2 is diminished, the efficiency is increased, as shown by equation (121).

If $v = 10$ feet per second (about 6.8 miles per hour), we find

v_2	U	k	E
10	$0. k \dfrac{\delta}{g}$	x	1.00
15	$750 k$ "	120.0	0.80
20	$2000 k$ "	45.0	0.67
30	$6000 k$ "	15.0	0.50
40	$12000 k$ "	7.5	0.40
100	$90000 k$ "	1.0	0.16

4

The sections k here given are for equal works, U. If the velocity of exit be constant, then will the work increase directly as the area of the section while the efficiency remains the same. These are without frictional resistances.

The pressure against the side of the vessel opposite the orifice due to the reaction of the water will be found from equation (117) by dividing the work done by the space over which the work is done, or

$$P_1 = \frac{U_1}{v} = M v_2, \quad \quad \text{. . . (123)}$$

which is *the momentum of the jet per second relative to the orifice.*

To impart to the water taken up the uniform velocity v would require the constant pressure

$$P_2 = Mv ;$$

hence the resultant pressure producing work would be

$$P = P_1 - P_2 = M (v_2 - v), \quad \text{. . . . (124)}$$

and the resultant work would be

$$U_2 = M (v_2 - v)v, \quad \text{. . . . . (125)}$$

as before found in equation (120).

The speed of a jet propeller depends upon the form of the vessel and the nature of the fluid; but the pressures due to the action of the jet will be the same whether it issue into a vacuum, or into air or water, or a more viscous fluid. If a block be placed before the jet so close to the vessel as to obstruct the flow of water as a jet, the conditions will be changed, and the forward pressure will then be due partly to the direct pressure exerted by the pumps. If a piston, having a long piston-rod projecting against a firm body outside the vessel, be forced backward, the forward pressure, effective in driving the vessel, would be that exerted by the pumps less the frictional resistances.

The efficiency of the jet propeller as a motor is comparatively small in practice. This is due to the great loss of energy in the jet. The *entire* energy in the jet is lost. If the vessel be anchored, and the velocity of the jet be v_2, the pressure will be

$$P_1 = Mv_2,$$

the work will be

$$U = 0,$$

and the

$$energy\ lost = \tfrac{1}{2}Mv_2^2.$$

If the speed v of the vessel is small, then

$$P_1 = Mv_2,$$

$$U = Mv_2v,\ \text{nearly,}$$

$$energy\ lost = \tfrac{1}{2}Mv_2^2,\ \text{nearly,}$$

and the energy lost will generally exceed considerably the useful work.

39. *The difference of the moment of the momentum of the water on entering and leaving the wheel, equals the moment exerted by it on the wheel.* (Proof for the frictionless turbine by J. Lester Woodbridge, graduate of Stevens Institute, 1886.)

Consider the effect of water passing along a smooth, curved horizontal tube rotating about a vertical axis. Conceive the water to be divided into an infinite number of filaments by

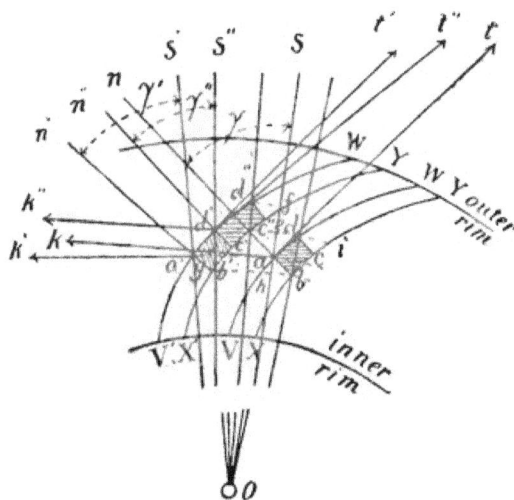

FIG. 18.

vanes similar to those of the wheel, but subjected to the condition that, at each point, their width, id, Fig. 18, measured on the arc, whose centre is O, shall subtend at the centre a constant angle $d\theta$. Conceive each filament to be divided into small prisms, whose bases are represented by the shaded areas $a'b'c'd'$, $d'c'c''d''$ and $abcd$, by vertical planes normal to the vanes making the divisions ac, ef, intercepted on the radius by circles passing through the consecutive vertices on the same vane, a', d', d'', etc., equal.

Let ρ = the radius vector;

x = the height of an elementary prism;

then, $d\rho = ac$, cf, etc.;

$\rho d\theta d\rho = abcd$, etc., = area of the base of an infinitesimal prism;

$x\rho d\theta d\rho$ = volume of an infinitesimal prism;

$x\delta\rho d\theta d\rho = m$ = the mass of prism, δ being its density or mass of unit volume;

$\gamma = san$ = angle between the normal to the vane at any point, and the radius $O\alpha$ prolonged through that point;

v = velocity of a particle along the vane at ρ, which is assumed to be the same in all the vanes at the same distance from the centre;

ω = the uniform angular velocity of the wheel, and

p = the pressure of the water at the point ρ due to a head, but not due to deflection.

Let ρ be the independent variable, and dt the time required for the element $a'b'c'd'$ to move its own length, $d\rho$, and aa' the distance passed through by this element circumferentially in the same time, dt, then

$$dt = \frac{d\rho}{v \sin \gamma'}$$

and,

$$aa' = \omega\rho dt = \omega\rho \frac{dt}{d\rho}d\rho.$$

The mass m will have two motions: one along the vane, the other with the wheel perpendicular to the radius. By changing its position successively in each of these directions, both its velocity with the wheel and its velocity along the vane may suffer changes both in *amount* and *direction*, as follows:

(I.) By moving from a to a', Fig. 18, in the arc of a circle—

(1.) $\omega\rho$ may be increased or diminished;

(2.) $\omega\rho$ may be changed in direction;

(3.) v may be increased or diminished;

(4.) v may be changed in direction.

(II.) By moving from a to a' along the vane—

(5.) $\omega\rho$ may be increased or diminished;

(6.) $\omega\rho$ may be changed in direction;

(7.) v may be increased or diminished;

(8.) v may be changed in direction.

These changes give rise to corresponding reactions, as follows:

(No. 1.) Since the element is to move from a to a' in the arc of a circle, $\omega\rho$ will be constant, and hence the reaction $= 0$.

(No. 2.) By moving from a to a', the velocity $\omega\rho$ is changed in direction from ak to $a'k'$ in the time dt. The momentum is $m\omega\rho$, and the rate of angular change is

$$\frac{kak'}{dt} = \frac{\omega dt}{dt} = \omega,$$

and hence the force upon the element producing motion in the arc of a circle will be radially inward and the reaction will be $m\omega^2\rho$ radially outward. This is generally called the *centrifugal force*, as designated by most writers. Resolving into two components, we have

$$m\omega^2\rho \sin \gamma \text{ along the vane,}$$
$$m\omega^2\rho \cos \gamma \text{ normal to the vane.}$$

(No. 3.) According to the conditions imposed, this value of v is the same at a' as at a, hence, for this case, the reaction will be zero.

(No. 4.) In moving from a to a' the velocity along the vane, v, is changed in direction from at to $a't'$ at the rate ω as in No. 2.

The momentum is mv, and the force will be $mv\omega$, which acts in the direction bn. Since the particle will be driven by the vane XY, and the reaction will be in the direction nb; which being resolved, gives

0 along the vane,

$- mv\omega$ normal to the vane.

(No. 5.) In passing from a' to d'; at d' the circular velocity will be greater than at a' by the amount

$$\omega d\rho,$$

and the acceleration will be

$$\omega \frac{d\rho}{dt},$$

requiring a force $m\omega \frac{d\rho}{dt}$ tangentially to the wheel in the direction of motion, the reaction of which will be

$$m\omega \frac{d\rho}{dt}$$

but backwards, and its components will be

$$m\omega \frac{d\rho}{dt} \cos \gamma \text{ along the vane,}$$

$$- m\omega \frac{d\rho}{dt} \sin \gamma \text{ normal to the vane.}$$

(No. 6.) In passing from a' to d', $\omega\rho$ will be changed in direction by the angle between $k'a'$ and $k''d'$, or

$$a'Od' = \frac{a'g}{\rho} = \frac{d\rho \cot \gamma}{\rho},$$

and the rate of angular change will be

$$\frac{\cot \gamma}{\rho} \cdot \frac{d\rho}{dt},$$

and the momentum being

$$m \omega \rho,$$

the reaction will be

$$m \omega \cot \gamma \frac{d\rho}{dt},$$

which acts radially inward and its components are

$$- m \omega \cos \gamma \frac{d\rho}{dt} \qquad \text{along the vane,}$$

$$- m \omega \cot \gamma \cos \gamma \frac{d\rho}{dt} \text{ normal to the vane.}$$

(No. 7.) By moving from a' to d, v will be increased by an amount

$$\frac{dv}{d\rho} d\rho,$$

in the time dt, and the reaction will be

$$m \frac{dv}{d\rho} \cdot \frac{d\rho}{dt},$$

which will be outward along the vane, and the reaction will be directly backward along the vane, and hence is

$$- m \frac{dv}{d\rho} \cdot \frac{d\rho}{dt} \qquad \text{along the vane,}$$

$$0 \quad \text{normal to the vane.}$$

(No. 8.) In passing from a' to d, v is changed in direction by two amounts: the angle γ changes an amount

$$d (\gamma'' - \gamma') = - \frac{d\gamma}{d\rho} d\rho.$$

This is negative, for a differential is the limiting value of the second state minus the first, and the first is here larger.

But this is not the total change, since γ'' is measured from a radius making an angle

$$\frac{d\rho \cot \gamma}{\rho},$$

with Oa' as in No. 6; hence the total change will be the sum of these, and the *rate* of change will be the sum divided by dt, which result, multiplied by the momentum mv, will give the reaction, which will be normal and in the direction $b'n'$ or

$$0 \qquad\qquad \text{along the vane,}$$

$$mv\left[\frac{\cot \gamma}{\rho} \cdot \frac{d\rho}{dt} - \frac{d\gamma}{d\rho} \cdot \frac{d\rho}{dt}\right], \text{ normal to the vane.}$$

This completes the reactions. Next consider the *pressure in the wheel*. The *intensity* of the pressure on the two sides ab and cd differs by an amount

$$dp = \frac{dp}{d\rho} d\rho.$$

The area of the face is $dc \times x = x\rho d\theta \sin \gamma$, and the force due to the difference of pressures will be

$$x\rho d\theta \sin \gamma \frac{dp}{d\rho} d\rho.$$

If dp is positive, which will be the case when the pressure on dc exceeds that on ab, the force acts backwards, and the preceding expression will be *minus* along the vane. In regard to the pressure normal to the vane, if a uniform pressure p existed from one end of the vane $V W$ to the other, the resultant effect would be zero, since the pressure in one direction on $V W$ would equal the opposite pressure on XY. If, however, in passing from d to a, the pressure increases by an amount $-dp$, since Va is longer than Xb, the pressure on Va will exceed that on Xb by an amount

$$- dp . x \times ah = - dp.x.\rho d\theta \cos \gamma = - x\rho \cos \gamma d\theta \frac{dp}{d\rho} d\rho.$$

Collecting these several reactions, we have

	NORMAL TO THE VANE.	ALONG THE VANE.
(1.)	0	0
(2.)	$+ m\omega^2\rho \cos \gamma.$	$+ m\omega^2\rho \sin \gamma.$
(4.)	$- m\omega v.$	$0.$
(5.)	$- m\omega \sin \gamma \dfrac{d\rho}{dt}.$	$+ m\omega \cos \gamma \dfrac{d\rho}{dt}.$
(6.)	$- m\omega \cot \gamma \cos \gamma \dfrac{d\rho}{dt}.$	$- m\omega \cos \gamma \dfrac{d\rho}{dt}.$
(7.)	0	$- m \dfrac{d\rho}{dt}.\dfrac{dv}{d\rho}.$
(8.)	$+ mv\left[\dfrac{\cot \gamma}{\rho} . \dfrac{d\rho}{dt} - \dfrac{d\gamma}{d\rho} . \dfrac{d\rho}{dt} \right].$	$0.$
(9.)	$- x\rho \cos \gamma \dfrac{dp}{d\rho} d\rho d\theta.$	$- x\rho \sin \gamma \dfrac{dp}{d\rho} d\rho d\theta$

The sum of the quantities in the second column, neglecting friction, will be zero ; hence

$$m\omega^2\rho \sin \gamma - m \frac{d\rho}{dt}.\frac{dv}{d\rho} - x\rho \sin \gamma \frac{dp}{d\rho} d\rho d\theta = 0. \quad (123)$$

Substituting

$$\frac{d\rho}{dt} = v \sin \gamma, \text{ and } x\rho d\theta d\rho = \frac{m}{\delta}$$

and dividing by $m \sin \gamma$, we have

$$\omega^2\rho d\rho - \frac{1}{\delta} dp = vdv. \quad . \quad . \quad . \quad . \quad (124)$$

Integrating,

$$\left[\tfrac{1}{2}\omega^2\rho^2 - \frac{p}{\delta}\right]_{\text{limit}}^{\text{limit}} = \left[\tfrac{1}{2}v^2\right]_{\text{limit}}^{\text{limit}}. \quad \cdots \quad (125)$$

The sum of the quantities in the first column gives the pressure normal to the vane, which, multiplied by $\rho \sin \gamma$, gives the moment. This done, we have

$$d^2M = mv \sin \rho \left\{ \begin{array}{l} \omega\gamma\left(\dfrac{\rho}{v}\omega\cos\gamma - 2\right) - \rho v \sin\gamma \dfrac{dy}{d\rho} \\[2mm] + v\cos\gamma - \rho\dfrac{\cos\gamma}{v\delta}\dfrac{dp}{d\rho} \end{array} \right\}$$

Putting $mv\sin\gamma = \dfrac{\delta Q}{2\pi g}\,d\rho/\theta$, where Q is the quantity of water flowing through the wheel per second, and integrating in reference to θ between 0 and 2π, we have

$$dM_1 = \frac{\delta Q}{g}\left[\omega\rho\left(\frac{\rho}{c}\omega\cos\gamma - 2\right) - \rho v\sin\gamma\frac{dy}{d\rho} + v\cos\gamma - \rho\frac{\cos\gamma}{v\delta}\frac{dp}{d\rho}\right]d\rho$$

Multiplying (124) by

$$\frac{\rho}{c}\cos\gamma,$$

we have

$$\frac{\omega^2\rho^2}{c}\cos\gamma\,d\rho - \frac{\rho\cos\gamma}{c\delta}\frac{dp}{d\rho}\,d\rho = \rho\cos\gamma\frac{dv}{d\rho}\,d\rho,$$

which substituted above gives

$$dM_1 = \frac{\delta Q}{g}\left[-2\omega\rho\,d\rho + \rho\cos\gamma\frac{dv}{d\rho}\,d\rho + v\cos\gamma\,d\rho - \rho v\sin\gamma\frac{dy}{d\rho}\,d\rho\right](126)$$

the integral of which is

$$M_1 = \delta Q\left[-\omega\rho^2 + \rho v\cos\gamma\right] \div g$$

$$= -\frac{\delta Q\rho}{g}\left[\omega\rho - v\cos\gamma\right]_{\text{limit.}}^{\text{limit.}} \quad \cdots \quad (127)$$

But $\omega \rho - v \cos \gamma$ is the circumferential velocity in space of the water at any point, and $\delta Q \rho [\omega \rho - v \cos \gamma]$ is the moment of the momentum; hence, integrating between limits for inner and outer rims, *the moment exerted by the water on the wheel equals the difference in its moment of momentum on entering and leaving wheel.*

Let the values of the variables at the entrance of the wheel be $\rho_1, \gamma_1, v_1, p_1$, and at exit $\rho_2, \gamma_2, v_2, p_2$.

Equations (125) and (127) become

$$\tfrac{1}{2} \omega^2 (\rho_1^2 - \rho_2^2) - \frac{p_1 - p_2}{\delta} = \tfrac{1}{2} (v_1^2 - v_2^2). \quad . \quad . \quad (128)$$

$$M = \delta Q [\omega (\rho_1^2 - \rho_2^2) - \rho_1 v_1 \cos \gamma_1 + \rho_2 v_2 \cos \gamma_2]. \quad (129)$$

$$U = M_1\omega = \frac{\delta Q}{g} \omega [\omega(\rho_1^2 - \rho_2^2) - \rho_1 v_1 \cos \gamma_1 + \rho_2 v_2 \cos \gamma_2], (130)$$

Equation (58) for a frictionless wheel in which $\mu_1 = \mu_2 = 0$, reduces to equation (130). This principle simplifies the solution of certain special cases. Thus, in the Barker Mill, page 44, the momentum of the water entering the wheel will be zero, but of exit will be

$$M V_2.$$

where V_2 is the velocity of exit, relative to the earth, perpendicular to the arm, and the moment will be

$$M V_2 r_2 ;$$
$$\therefore U = M V_2 \cdot r_2 \omega$$
$$= Ps,$$

where P is the effective pressure on the arm opposite the orifice of the jet, and s the space passed over by the orifice in a second of time. If P_1 be the pressure on the arm due to the reaction, $M V_2$, of the jet, and P_2 the pressure which imparts to the water a momentum $Mr_2\omega$, then $P = P_1 - P_2$.

But

$$V = v_2 - r_2 \, \omega;$$

$$M = \frac{\delta \, Q}{g} = \frac{\delta \, k_2 \, v_2}{g};$$

$$(1 + \mu) \, v_2^2 = 2 \, g \, H + \omega^2 \, r_2^2;$$

$$\therefore \; U = \frac{\delta \, k_2 \, v_2}{g} \, (v_2 - r_2 \, \omega) \, r_2 \, \omega \tag{132}$$

$$= \frac{\delta \, Q}{g} \left(- r_2 \, \omega + \sqrt{\frac{2 \, g \, H + \omega^2 \, r_2^2}{1 + \mu}} \right) r_2 \, \omega,$$

as in equation (105).

40. Again, if the water quits the wheel radially, then the moment of the momentum of the quitting water will be zero, and

$$U = M \, V \, r_1 \, \cos \alpha \cdot \omega.$$

But

$$V \cos \alpha = V_1.$$

the tangential component of the velocity, or velocity of whirl ;

$$\therefore \; U = M \, V_1 \, r_1 \, \omega. \tag{133}$$

41. *In the frictionless Rankine wheel* the velocity of whirl equals the velocity of the initial rim of the wheel.

$$\therefore \; V_1 = r_1 \, \omega ;$$

$$\therefore \; U = M \, r_1^2 \, \omega^2. \tag{134}$$

The work will also equal the potential energy of the water, $W \, H = \delta \, Q \, H$, less the kinetic energy of the quitting water, $\frac{1}{2} \, M \, V_2^2$ (less the energy lost in resistances, $\mu \, v_2^2$, which in this case we neglect) ;

$$\therefore \; U = \delta \, Q \, H - \frac{1}{2} \, M \, V_2^2,$$

and since the water is assumed to quit radially

$$V_2 = r_2 \, \omega \cdot \tan \gamma_2 = r_1 \, \omega \tan \alpha. \tag{135}$$

The three preceding equations give

$$r_1\,\omega = \sqrt{\dfrac{2\,g\,H}{2 + tan^2\,\alpha}},$$

as in equation (76).

42. *Again, if the crowns are parallel discs* and the initial element of the bucket is radial, and if the water quits the wheel radially, and if the velocity of whirl equals the velocity of the initial rim, we have

$$U = M\,r_1^2\,\omega^2, \tag{136}$$

as in equation (134). But γ_2 will not be the same as in (135). To find it we have, neglecting the thickness of the walls of the buckets,

$$2\,\pi\,r_1\,c_1 = 2\,\pi\,r_2\,sin\,\gamma_2 \cdot c_2,$$
$$c_1 = V\,sin\,\alpha$$
$$r_1\,\omega = V\,cos\,\alpha$$
$$V_2 = r_2\,\omega\,tan\,\gamma_2$$
$$c_2 = r_2\,\omega;$$
$$\therefore tan\,\gamma_2 = \dfrac{r_1^2\,tan\,\alpha}{\sqrt{r_2^4 - r_1^4\,tan^2\,\alpha}}; \tag{137}$$

$$\therefore U = \delta\,Q\,H - \tfrac{1}{2}\,M\,V_2^2$$
$$= \delta\,Q\,H - \tfrac{1}{2}\,M\,r_2^2\,\omega^2\,\dfrac{r_1^4\,tan^2\,\alpha}{r_2^4 - r_1^4\,tan^2\,\alpha}. \tag{138}$$

Equations (136) and (138) give

$$r_1\,\omega = \sqrt{\dfrac{2\,g\,H}{2 + \dfrac{r_1^2\,r_2^2\,tan^2\,\alpha}{r_2^4 - r_1^4\,tan^2\,\alpha}}}. \tag{139}$$

Practical and Experimental Data and Results.

A STUDY OF THE TREMONT TURBINE.

(Revision of a paper by the Author in Vol. XVI. of Trans. of Am. Soc. of Mech. Engineers.)

44. The Tremont turbine furnishes an excellent example for testing the theoretical formulas for the proportions and deliverance of turbines. The one here analyzed was made by J. B. Francis, engineer, after the general pattern of the celebrated U. A. Boyden, Esq., turbines, which yielded, from careful experiments, an admitted efficiency of 88 per cent. The dimensions of a Tremont turbine and the careful and somewhat exhaustive efficiency tests are fully set forth by Francis, in his work entitled *Lowell Hydraulic Experiments.* (Fig. 19.)

The workmanship on these wheels was of high grade. The crowns, which were of cast iron, were accurately turned in a lathe, and the partitions (or walls) of the buckets were of Russia sheet-iron plates, $\frac{9}{64}$ of an inch thick. These plates fitted into grooves carefully cut in the crowns, on which were tongues projecting through mortices in the crowns, and the former headed down, thus securing the crowns to each other without bolts or rods, and forming smooth, unobstructed passages for the flow of water. In the Boyden wheel a "diffuser" was added, which consisted of two crown-like pieces outside the wheel, but not rotating with it, the space between which was diverging, producing a diminution of velocity of the escaping water, causing more of its energy to be imparted to the wheel. This device, it is said, added some 2 or 3 per cent. to the efficiency of the wheel. The Tremont turbine was not provided with this device. (Fig. 30.)

We have selected for analysis the experiment by Francis which yielded the highest efficiency; for our turbine formulas are strictly applicable only when the wheel is so proportioned and run as to give a maximum efficiency.

RESULTS OF EXPERIMENTS WITH THE TREMONT TURBINE.

Number of the experiment. (1)	Height of the regulating gate in inches. (4)	Number of revolutions of the wheel per second. $\frac{n}{60}$ (8)	Useful effect, or the friction of the brake in pounds avoirdupois raised one foot per second. U (10)	Total fall acting upon the wheel in feet. H (13)	Quantity of water which passed the weir, in cubic feet per second. Q (12)	Total power of the water in pounds avoirdupois raised one foot per second. w (16)	Ratio of the useful effect to the power expended. E (17)	Velocity due to the fall acting on the wheel in feet per second. $\sqrt{}$ (18)	Velocity of the interior circumference of the wheel in feet per second. $r_1\omega$ (19)	Ratio of the velocity of the interior circumference of the wheel to the velocity due to the fall acting on the wheel. (20)	Direction of the water leaving the wheel as indicated by the vane. (21) deg.	m.
4	11.49	1.59651	33348.3	12.554	156.6470	122663.3	0.27187	28.4169	33.8553	1.19138	18	3
6		1.46149	51680.6	12.653	152.9682	120174.0	0.43005	28.5287	30.9921	1.08635	22	56
8		1.39933	66845.5	12.720	147.2942	116864.8	0.57169	28.6041	27.7653	0.97067	29	49
10		1.18460	77154.6	12.800	144.8734	115667.0	0.66704	28.6470	25.1203	0.87546	35	37
13		1.23404	0.	12.510	163.4313	127527.3	0.	28.7366	37.8319	1.33366	41	26
14		1.06744	83969.8	12.856	162.5180	144384.2	0.73475	28.7946	22.6559	0.78716	46	18
16		0.99945	86314.4	12.880	141.9762	141450.8	0.75614	28.5835	21.1940	0.73604	48	26
18		0.94507	87392.7	12.886	140.4657	112848.6	0.77442	28.7902	20.0409	0.69026		
20		0.91116	87819.8	12.896	140.0096	112532.3	0.78040	28.5736	19.3219	0.67113		
22		0.89713	89076.2	12.429	139.6674	112564.5	0.78384	28.3750	19.0243	0.66048		
23		1.78994	0.	12.499	139.0291	123553.6	0.	28.2750	37.7662	1.33367	50	37
25		0.88496	84199.9	12.503	161.6944	111559.4	0.78410	28.8047	18.7662	0.65150	58	10
30		0.85106	88278.9		138.3283	111218.1	0.78375	28.8092	18.0474	0.62645	61	54
31	11.48	1.78971	0.	12.915	162.2668	130950.2	0.	28.8225	37.9621	1.33635		
32		0.83624	88220.8	12.941	138.0860	111584.0	0.79294	28.8515	18.4474	0.61525	96	12
34		0.78101	87942.8	12.989	137.7976	111463.0	0.78863	28.8493	17.7530	0.57624	99	25
35		0.74211	82096.5			111380.7	0.78340		16.6251	0.54549		

18	131										
45	139	0.47150	13.6922	28.8504	0.76978	109077.1	155.1415	12.940	83965.5	0.64568	11.48	37
25	147	1.33344	37.8674	28.3557	0.	126071.1	161.6944	12.500	0.	1.78871	...	38
		0.44045	12.7235	28.8916	0.73277	109077.0	134.7976	12.973	82108.8	0.60000	...	40
		0.39071	11.2889	28.6906	0.72499	108265.8	133.7528	12.077	78492.2	0.53332	...	41
52	60		0.	28.6322	0.	106280.4	133.6636	12.797	0.	0.	...	43
27	66	1.33521	37.8168	28.3660	0.	126034.8	132.0237	12.934	0.	1.78333	...	45
44	89	0.61126	17.6447	28.8593	0.79104	112009.3	138.6244	12.948	88003.9	0.83205	...	46
0	12	0.59120	17.0327	28.8638	0.78904	111832.0	138.4360	12.952	88240.1	0.80321	...	48
19	28	0.56492	16.3058	28.6468	0.78723	111613.5	138.1559	12.949	87895.8	0.70893	8.55	50
56	28	1.09802	31.4548	28.8159	0.34389	114060.7	143.8319	12.938	83452.4	1.48331	...	51
47	47	0.84023	24.2121	28.8616	0.67402	110917.6	137.7518	12.999	74827.2	1.14177	...	53
30	59	0.73281	21.7272	28.8783	0.72485	110664.7	137.0026	12.950	80215.4	1.02459	...	55
39	94	0.68344	19.7365	28.9161	0.74977	109232.2	135.0874	12.960	81911.6	0.93871	...	56
4		0.63463	18.3611	28.9461	0.76449	109082.5	133.3014	12.999	82195.5	0.88528	...	58
		0.58922	17.0556	28.9484	0.76260	107246.3	131.9960	13.026	81786.2	0.89429	...	61
30	6	0.51848	15.0051	28.9484	0.75012	106366.6	128.9869	13.086	79838.4	0.79679	...	63
32	12	1.27570	36.4332	28.6041	0.	118652.1	149.5470	12.720	76785.1	1.71808	...	64
17	91	0.85363	32.8262	29.1657	0.	100194.6	128.9985	13.170	0.	1.54799	5.65	65
27	39	0.85216	29.6764	29.0028	0.30046	94399.3	123.3511	13.077	29054.7	1.35135	...	66
56	69	0.81833	23.9997	29.1125	0.46592	93804.9	114.2369	13.176	53142.4	1.08400	...	68
52	144	0.70628	19.5831	29.19.3	0.45763	92189.4	111.5197	13.253	58725.1	0.92348	...	70
28	55	0.76979	16.4866	29.2611	0.45550	89707.1	108.0452	13.311	58776.2	0.77482	...	72
30	74	0.49707	14.5530	29.2776	0.64127	87720.9	105.5341	13.396	56253.1	0.68627	...	74
32	0	0.32906	9.6830	29.3719	0.54757	84110.0	100.5410	13.412	46056.1	0.45662	...	76
9	4	0.62061	18.1246	28.7868	0.78100	106360.8	137.3618	12.883	86007.6	0.85470	9.06	77
40	90	0.58669	16.7351	28.8192	0.78350	109973.0	136.5469	12.912	86164.4	0.78918	...	79
	81	0.90596	26.4865	29.3006	0.33204	66879.0	80.4534	13.347	0.	1.24002	...	80
		0.69369	20.3622	29.3533	0.46314	64018.1	76.6213	13.395	21256.0	0.99022	2.875	82
		0.48390	14.2480	29.441	0.42661	60124.6	71.8730	13.428	27985.1	0.67189	...	84
		0.33208	9.8061	29.3292	0.	57342.0	67.8158	13.556	24462.9	0.46343	...	86
			0.	29.4856	0.	50896.6	60.3593	13.516	0.	0.	...	88
		0.43806	13.1386	29.9993	0.14483	38340.8	38.2210	13.986	4998.8	0.61958	1.00	90
		0.48677	14.6979	30.0099	0.10176	33683.5	38.6599	14.001	3427.7	0.68896	...	91
		0.27570	8.2195	30.0503	0.24042	32508.0	37.1733	14.020	7815.6	0.38760	...	92

The coefficients μ_1 and μ_2 are uncertain quantities. Weisbach considers $\mu_1 = 0.10$, and μ_2 from .05 to 0.10. We will assume $\mu_1 = 0.10$, and for the assumed Boyden wheel $\mu_2 = 0.05$. Our analysis will show that $\mu_2 = 0.15$ nearly, for the Francis wheel.

For the wheel we now analyze we have (see *Lowell Hyd. Exp.*),

$r_1 = 3.375$ feet ; $r_2 = 4.146$ feet.

$\alpha = 21°$) as measured from Plate III. of *Experiments.**
$\gamma_2 = 10°$ } (We have used $\gamma_2 = 13°$ and $15°$ for reasons
$\gamma_1 = 90°$) which appear later.)

$Q = 138.2$ cubic feet.

$H = 12.9$ feet.

$h_1 = 15.11$ feet.

$h_2 = 2.2$ feet.

$U = 111,218.1$, total power of the water in foot-pounds per second.

$E = 0.79375$ efficiency.

$H.P. = 161$, about, as measured by the brake.

$n = 0.85106$, number of revolutions per second.

$y_1 = 0.9368$ feet, depth between the inner edges of the crowns.

$y_2 = 0.9314$ feet, depth between the outer edges of the crowns.

$N = 44$, number of buckets.

$g = 32.16$.

With this data we solve as follows :

$$Rev. \ per \ sec. = \frac{\omega}{2\pi}. \tag{140}$$

* Rodmer in his work gives $\alpha = 28°$, $\gamma_1 = 90°$, $\gamma_2 = 22°$, but these do not agree with the plates given by Francis, neither do they produce results agreeing with *Experiments*. Rodmer considers these as *mean* angles of the stream.

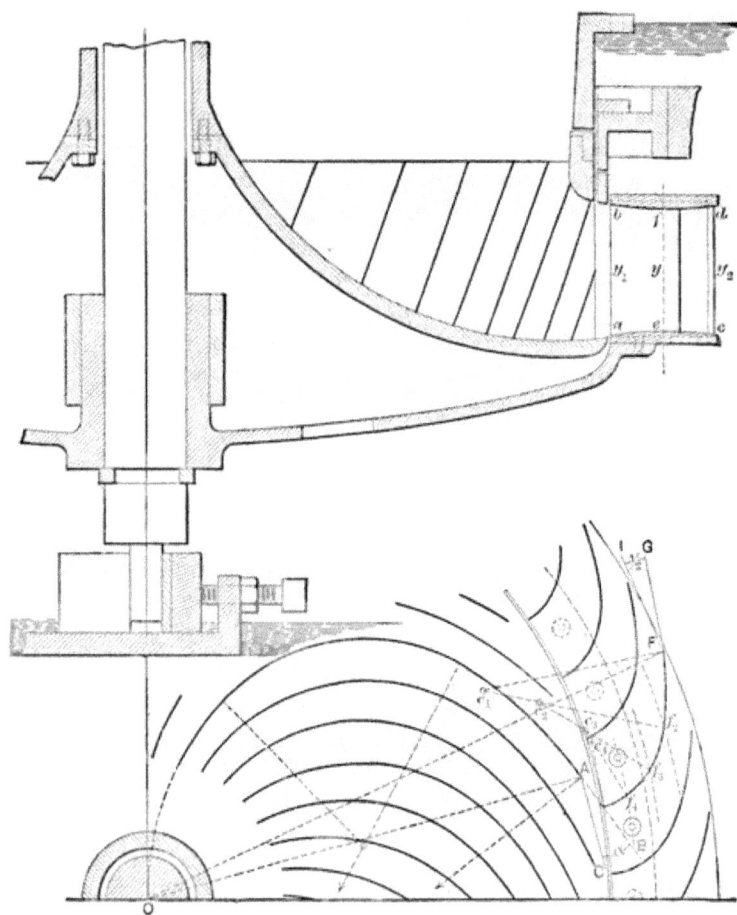

Fig. 19.

TABLE IX.

	$a = 21°$ $\gamma_2 = 13°$ $\mu_2 = 0.20$	$a = 21°$ $\gamma_2 = 13°$ $\mu_2 = 0.10$	$a = 21°$ $\gamma_2 = 13°$ $\mu_2 = 0.05$	Measured Values.
ω. Eq. (16)...........	5.369	5.476	5.514	......
Rev. per second........	0.855	0.872	0.879	0.851
E_{max} per cent. Eq. (16a)	79.308	84.84	87.76	79.375
v_1 feet. Eq. (18).......	6.968	7.102	7.157	......
v_2 feet. Eq. (19).......	23.115	23.989	24.525	......
$\frac{v_2}{v_1} = \frac{k_1}{k_2} =$ Eq. (20)....	3.316	3.37	3.426	......
y_1 feet. Eq. (21)......	0.957	0.939	0.932	0.9368
y_2 feet. Eq. (21).......	1.103	1.063	1.039	0.9314

It will be observed that for $\mu_2 = 0.05$ the efficiency is over 87 per cent., and if a diffuser would add 2 per cent., we would have an efficiency exceeding that admitted for the Boyden wheel. But it seems improbable that the prejudicial resistance can be so low; and it is well to observe that for $\alpha = 20$ degrees, $\mu_1 = 0.10 = \mu_2$, and $\gamma_2 = 10$ degrees, a theoretical efficiency nearer 90 per cent. will be found.

It will be seen that for $\mu_2 = 0.20$, some of the numbers given in the first column of the preceding table agree well with actual values. Thus the revolutions are:

Computed . . . . 51.3 per minute.

Measured . . . . 51.1 per minute.

Also the efficiency is:

Computed . . . . 79.308 per cent.

From measurements . . . . . . 79.375 per cent.

The initial depth between the crowns is:

Computed . . . . 0.957 feet.

Measured . . . . 0.9368 feet.

Difference . . . . .0202 feet.

The outer depths between the crowns are :

Computed . . . . 1.103 feet.
Measured . . . . .9314 feet.

Difference . . . . .1716 feet.
or . . . . 2.059 inches.

Not only do the computed depths exceed the measured ones,
but our computed depth is less for the inner rim than for the
outer, while the measured values are the reverse. This cannot
be dismissed with the remark that the wheel was improperly
proportioned ; for the 138.2 cubic feet per second passed the
weir : therefore this requires further investigation.

45. *The depths between the crowns* were determined as follows :
Dropping the subscripts in the notation and placing ρ for r_1 or r_2,
we have

$$y (2 \pi \rho - Nt \div \sin \gamma) \, r \sin \gamma = Q \qquad (141)$$

in which $2 \pi \rho =$ whole circumference at distance ρ from centre.
$Nt =$ thickness of N bucket plates.
$Nt \div \sin \gamma =$ thickness measured on an arc.
$y (2 \pi \rho - Nt \div \sin \gamma) =$ free ring section for passage of
water.
$r \sin \gamma =$ radial component of the velocity in the bucket, and
the product gives the quantity of flow in feet per second, or Q.
Hence, for initial and terminable values, we have, since Q is
constant,

$$\frac{y_2}{y_1} = \frac{v_1 (2 \pi r_1 \sin \gamma_1 - Nt)}{v_2 (2 \pi r_2 \sin \gamma_2 - Nt)} = \frac{v_1 \left(2 \pi \times 3.38 - \dfrac{44 \times 9}{12 \times 64}\right)}{v_2 \left(2 \pi r_2 \sin \gamma_2 - \dfrac{44 \times 9}{12 \times 64}\right)} \quad (142)$$

If the thickness of the bucket plates be neglected, we have
simply (γ_1 being 90 degrees),

$$\frac{y_2}{y_1} = \frac{r_1}{r_2} \cdot \frac{v_1}{v_2 \sin \gamma_2},$$

which for $\gamma_2 = 13$ degrees gives

$$\frac{3.38}{4.20} \cdot \frac{6.968}{23.115 \times 0.245} = 1 \text{ very nearly},$$

or the depths would be equal; and that the outer depth shall be less than the inner, γ_2 must exceed 13 degrees. We tried 15 degrees and made other computations, as follows:

TABLE X.

	$Q = 138.2$ $a = 21°$ $\gamma_2 = 15°$ $\mu_2 = 0.20$	$Q = 135.5$ $a = 21°$ $\gamma_2 = 15°$ $\mu_2 = 0.15$	$Q = 138.2$ $a = 23°$ $\gamma_2 = 14°$ $\mu_2 = 0.20$	$Q = 138.2$ $a = 23°$ $\gamma_2 = 15°$ $\mu_2 = 0.20$	Measured.
ω.....................	5.359	5.407	5.364	5.346	......
Rev. per second......	0.853	0.867	0.854	0.851	.851
E_{max} per cent........	78.33	80.9	78.90	78.10	79.375
r_1 feet	6.954	7.017	7.696	7.67	......
r_2 feet	23.048	23.573	23.15	23.19	......
y_1 feet.............	0.9439	0.9317	0.862	0.870	.9368
y_2 feet.............	0.9346	0.9104	1.00	0.945	.9314

In the first column the buckets are still too deep and the efficiency too small. Since the gate does not fit water-tight, a part of the 138.2 cubic feet which passed over the weir may have escaped at the gate, and hence the entire quantity may not have done work on the wheel. If there be $\frac{1}{8}$ of an inch clearance at the gate (this assumption is gratuitous, but will serve for illustration) there will be an annular opening of $2 \times 3\frac{1}{4} \times 3.38 \times \frac{1}{8} \times \frac{1}{12} = 0.222$ square feet. It will be found hereafter that the internal pressure at the gate is 2648 pounds per square foot, and the atmosphere and 2.2 feet head in the wheel pit gives an outside pressure of 2254 pounds, leaving 394 pounds inside effective pressure at that point, which would produce

$$a\ velocity = \sqrt{2\,g \times \frac{394}{62.2}} = 20 \text{ feet nearly};$$

hence the volume of water which would escape under these con-

ditions would be $0.62 \times 0.222 \times 20 = 2.75$ cubic feet nearly,
allowing 0.62 for the coefficient of discharge. This would leave
135.5 cubic feet of water to do work on the wheel. The leakage
produces a slight discontinuity.

The computed (hydraulic) efficiency ought to equal the brake
efficiency plus that of the shaft friction plus that lost by leakage;
so another computation was made, using

$$Q = 135.5, \; a = 21°, \; \gamma_2 = 15°, \; \mu_1 = 0.10, \; \mu_2 = 0.15,$$

the results of which are in the second column of Table X., and
are very good. The revolutions are *one* more per minute than
those observed; the efficiency, 1.525 per cent. greater than that
observed, which difference is probably about right; the inner
depth, y_1, is almost exact, but y_2 is perceptably less than that
measured. A reduction of 2 per cent. of the speed to reduce it
to that observed would reduce the efficiencies but slightly, as
shown by the experiments on page 68. Thus, in experiment 32
the speed is about 2 per cent. less than in No. 30, which gave
the maximum, but the efficiency is reduced from 0.79375 only
to 0.79294. In No. 48 the speed is nearly 6 per cent. less, but
the efficiency is reduced about 0.7 per cent., or to 0.789. It will
be observed that the wheel delivers nearly the maximum work
when the speed is within 5 or 6 per cent. above or below that
which gives the maximum efficiency.

There is some uncertainty about the correct measure of the
section of discharge. In equation (141) $\dfrac{t}{\sin \gamma}$ gives the thickness of a bucket-wall measured on the arc whose radius is ρ, on
the hypothesis that the arc is a straight line, which is sufficiently
accurate when the wall is very thin as in this case. Also $r \sin \gamma$
is assumed to be constant for all points in a ring section which is
accurate for a fillet, and sufficiently so for a very narrow stream.
But these conditions are changed when applied to finite streams,
as in actual wheels. This will be more clearly seen if equation

(141) is so transformed as to apply to a single bucket, for which it becomes at the terminal end

$$y_{2} \, N \left[\frac{2 \, \pi r_{2}}{N} \sin \gamma_{2} - t \right] v_{2} = Q, \qquad (142)$$

in which $\frac{2 \, \pi r_{2}}{N} = af$, Fig. 20, the distance measured on the arc covered by one bucket and partition-wall.

$t = he$, the thickness of the partition-wall.

Francis assumed that the correct section of discharge was not the arc ad, but was the least section, and experiments and computations combined confirm, at least approximately, the correctness of this assumption. Hence with a as a centre find by trial an arc which will be tangent at h—draw ah, it will be normal to dh, and will be the base of the least section of flow. Through the middle point of ah at g pass the middle arc gt of the bucket. Then considering g as the terminus of the middle fillet, find γ, r, and v for that point, and call that velocity the mean, and let v_{m} be the mean velocity; then $v_{\mathrm{m}} \, ah \, . \, y \, . \, N = Q$. In this case γ at g will exceed γ_{2} at a, r at g will be less than r_{2} at a and v_{m} will be less than v_{2}. We have not determined these quantities at g, for our knowledge of what takes place in the section ah is not definite, for it is a state bordering on discontinuity, since the stream follows the bucket to that section, and from that flies off into space in the direction $nahm$ more or less radial. We have not positive means of determining whether this view is correct, for we do not know if the bucket was filled with a live stream from end to end; but we assume that it was, and are seeking conditions which will be consistent with this assumption. We have already seen that by increasing γ_{2} from 13°, the vane angle, to 15° gives results which agree closely with our assumption, and we anticipate that where the terminal angle γ_{2} is small and the buckets numerous a similar result will follow in other cases.

Prolong *ah* to *c*, and make *cq* and *hi* perpendicular to *ah* ; *aj* tangent at *a* intersecting *cq* at *j* ; then

$$aj \sin aje = ae,$$
$$ah = ae - t \text{ (very nearly).}$$

Drawing *ab'* tangent to the arc of the bucket at *a*, then $b'aj = \gamma_2$, and *aje* will exceed γ_2, while *aj* will be less than the arc *af*, so that it may require a computation to determine whether *af* sin $\gamma_2 - t$ is greater or less than the least measured distance, *ah*. In Fig. 20, γ_2 is 25°; in the Tremont, 10°. The computed distance will be

$$b = \frac{2 \pi \times 4}{44} \sin 10° - \frac{9}{12 \times 64} = 0.1095 \text{ feet.}$$

The least measured distance *ah* = 0.1875 feet. The computed velocity, v_2, will exceed the mean velocity, and *b*, being less than the least base, there is a partial compensation in the product $b\, v_2$, so that $y_2 = Q \div Nb v_2$ approximates toward the correct value. According to our analysis,

$$y_2\, v_2\, (af \sin 13° - t) = \frac{Q}{N}$$

gave too large a value for y_2 ; then we found by trial that

$$y_2\, v_2\, (af \sin 15° - t) = \frac{Q}{N}$$

gave a value for y_2 a little too small. Had the mean velocity, v_m, been found, it would have been necessary to increase γ still more, that we may have the equality

$$y_2\, v_m\, (af \sin \gamma - t) = \frac{Q}{N}.$$

But in this case y_2 would be a mean depth along *ah*. If the bucket be divided into very small imaginary ones, say *n* such, we have

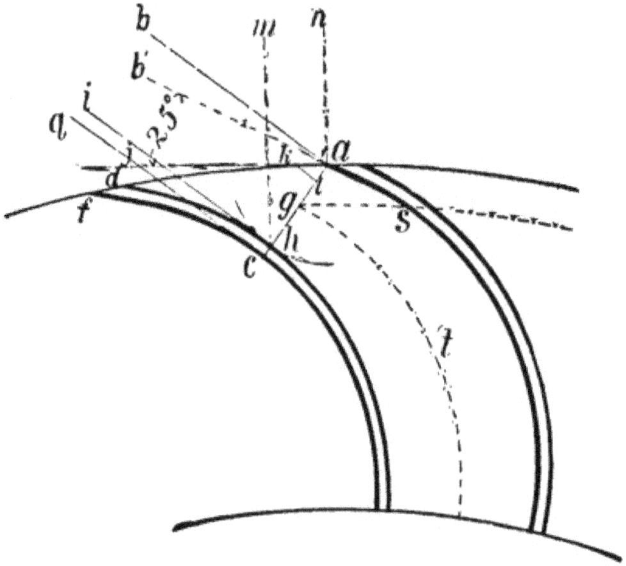

Fig. 20.

$$y_2\, v_2\ \frac{ad}{n}\ \text{sin}\ \gamma_2 = \frac{Q}{n\,.V},$$

but this is equivalent to assuming that the velocity is uniform throughout the cross-section and equal to v_2, which is true only for very narrow streams. Cancelling n will reduce the equation to the same as (141).

From the results in Tables IX. and X., we conclude that the prejudicial resistance μ_2 is less than 0.15 and exceeds 0.10 in this wheel. The terminal angle of the guide α perhaps ought to be increased in the analysis, for the same reason as that applied to γ_2, but not to the same amount, for the pressure at the gate will cause the water to follow the back of the guide more nearly to its end.

We try another computation, making

$$Q = 136,\ \alpha = 22°,\ \gamma_2 = 15,\ \mu_1 = \mu_2 = 0.13,$$

with the following results :

$Rev. = 0.862$; $E = 0.821$; $y_1 = 0.884$, $y_2 = 1.038$, which results are not as good as those previously found. Since this value of y_2 exceeds y_1, γ_2 ought to exceed 15°. Making

$$\gamma_2 = 17°,\ \mu_1 = 0.10 = \mu_2,$$

and other quantities as before, we find

$$Rev. = 880;\ E = 0.8361;\ y_1 = 0.872,\ y_2 = 0.869,$$

where the ratio of $y_1 \div y_2$ is good, but the other values are too large.

In regard to the value of α, measurements on Francis plates (Lowell *Hydraulic Experiments*) show that α, as it is measured, will not exceed 22°, and yet Rodmer gives the mean angle α as 28°. We solve the problem with Rodmer's values $\gamma_2 = 22°$, $\alpha = 28°$, assuming $\mu_1 = 0.10$, $\mu_2 = 0.05$; also $\mu_1 = 0.10 = \mu_2$, with the following results :

TABLE XI.

$$r_1 = 3.38 \text{ ft.}, \ r_2 = 4.0 \text{ ft.}, \ \gamma_1 = 90°, \ \alpha = 28°,$$
$$\gamma_2 = 22°, \ H = 12.9 \text{ ft.}, \ Q = 136 \text{ cu. ft.}$$

	$\mu_1 = 0.10$ $\mu_2 = 0.05$	$\mu_1 = 0.10$ $\mu_2 = 0.10$
Rev. per second.................	0.869	0.8636
E_{max}, per cent.................	81.96	79.30
r_1 feet per second.................	9.822	9.755
r_2 feet per second.................	23.525	23.059
V.................................	20.9012	20.770
y_1 Eq. (141).................	0.668	0.673
y_2 Eq. (141).................	0.650	0.663
y_1/y_2.................................	1.027	1.015

The brake efficiency of this wheel being 0.79375, its hydraulic efficiency will probably be 81 or 82 per cent., so that the last hypothesis above is not admissible. Rodner's values for α and γ_2 with low resistances give a satisfactory result for the efficiency, and about 1 revolution more per minute than the measured one; but they do not produce the proper values for the depths y_1 and y_2.

The total least section of the buckets is

$$k'_2 = 7.687 \text{ sq. ft.},$$

and if the velocity be $v_2 = 23.5$, the depth would be

$$y_2 = \frac{136}{23.5 \times 7.687 \text{ sq. ft.}} = 0.753 \text{ ft.},$$

so that the proper depth cannot be found with this data, and it would require a coefficient of contraction of 0.76 to give 0.93 +.

The total least section of the guides, as measured, was $K'' = 6.5371$ sq. ft.; the total initial arc of the buckets was $k'_1 = 19.380$ sq. ft.

To show the effect of a large terminal angle of the bucket, a computation was made with the data

$r_1 = 3.38$, $r_2 = 4.00$, $\alpha = 22°$, $\gamma_1 = 90°$, $\mu_1 = \mu_2 = 0.10$, and $\gamma_2 = 28°$, and the result gave

$$E_{max.} = 0.738, \quad Rev. = 0.850,$$

an efficiency nearly 6 per cent. less than the observed, at about the same speed. With the same data, except $\mu_1 = 0.20$, $\mu_2 = 0.05$, it was found:

$$E_{max.} = 0.780, \quad Rev. = 0.820,$$

so that with less revolutions the efficiency approximated to the actual.

As seen, the method of "trial vane angles" may be made to fit the conditions of the wheel when the results are known; but as a method of design it is not satisfactory, if not worthless. The "least section" as used by Francis is better, but the conditions of that cannot be realized, for the stream cannot change suddenly from the tangent $a b'$ to $a b$. We now suggest the following, which, though not perfect, offers the most promising condition for a practical solution. Draw the tangent $a b'$, Fig. 20a, and through n, the middle of the arc $a d$, erect a perpendicular $p r$ to $a b'$; it will be found that in this case p is the point of tangency to $p d$ of a line parallel to $a b'$. A particle at a would, if free, pass off in the direction $a b'$ tangent to the bucket, and would trace that tangent on a plane rotating with the crowns. Similarly, a particle at h would, if free, go off in a tangent at h; but this tangent is not parallel to $a b'$. Similarly at all points in $a h$ the particles would, if free, move in tangential directions to their paths; but those tangents all have varying directions. The particles, however, are not free. Those moving along the back of the bucket tend continually to leave it, and, with other particles in $a h$, force the stream toward $a b'$, and the stream may leave the back of the bucket before reaching p, while the mass of water in the vicinity of a opposes such deflection, the result being a contraction of the stream. If the bucket be full at exit, the

coefficient of contraction may be found. The middle line of
the bucket will pass through u, and at u the terminal angle γ_2
will be the same as at a ($10°$). With the data,

   $r_1 = 3.375$, $r_2 = 4.146$ (which call 4.1, since the buckets do
      not extend to the outer rim);

   $a = 21°$, $\gamma_1 = 90°$, $\gamma_2 = 10°$;

   $\mu_1 = 0.10$, $\mu_2 = 0.20$, $Q = 136$,

we recompute, finding

   Rev. per sec. 0.867.

   $E_{\max.}$ per cent. 81.18.

   c_1, feet 7.072.

   c_2, feet 22.57.

   y_1, feet 0.93.

   y_2 by equation (141) would exceed 1.3; but we will find it
      as follows:

Measure $p\,r$ on the drawing, finding

$$p\,r = 0.166,$$

and assume that the value of $c_2 = 22.57$ is the computed mean
velocity of the section, and assume (for trial) a coefficient of
contraction of 0.88; then

$$c_2 \times 0.88\ (44 \times 0.166)\ y_2 = 136;$$
$$\therefore\ y_2 = 0.9304 \text{ feet},$$
$$= 0.93 \text{ to the nearest hundredth},$$

which is not only sufficiently exact, but, in our ignorance of the
depth of the stream, may be actually too large; hence the coefficient 0.88 is sufficiently small. If the coefficient of discharge
be 0.90 the computed depth will be 0.9166 feet, which is about
$\frac{1}{8}$ of an inch less than the measured depth between the crowns,
and this may possibly be greater than the depth of the live
stream at that point.

The exact character of the stream in the buckets is unknown, especially at the terminus, so that perhaps the computed depth ought not to agree exactly with that of the wheel, but it is very certain that the initial end of the bucket was full, with gate fully open, for with the least closing of the gate there was a diminution in the volume discharged.

Many other computations have been made, but they add little if anything to our knowledge. As a result of this study it is inferred that for an outflow turbine similar to that of the Francis, the efficiency may be determined quite accurately by equation (16a), using the dimensions of the wheel, including the guide and bucket angles; and the initial depth between crowns to deliver a given amount of work may be determined by the second of equations (20); but the terminal depth is more accurately determined by means of a measurement across the stream, as shown above. It will be seen hereafter that the same general principles apply to other turbines.

No motor is designed which secures exactly a required result. There are minor elements which cannot be figured out exactly. After all the labor and study which has been given to the steam engine, it is necessary to "test" one to determine exactly what it will do, and it is the same with other motors. The performance of a well-proportioned turbine may, apparently, be determined as accurately by theory as that of any other motor.

46. In this wheel the inside of both the crowns is convex inward, as shown in the figure, the curvature being such as to reduce the depth between the crowns $\frac{3}{4}$ of an inch at $5\frac{1}{2}$ inches from the inner rim. This reduction is about $\frac{1}{15}$ of the initial depth. No reason is assigned for this form. If the object of this curvature was to conform to the form of the *vena contracta*, it appears to be too small. In any case, its effect will be to increase the velocity, and hence reduce the pressure in the wheel at that point; but if the internal pressure exceeds the ex-

ternal pressure (the external being that of an atmosphere 2116 pounds per foot, and the weight of 2.2 head of water in the wheel pit about 138 pounds, or a total of 2254 pounds) it seems to be unnecessary. The internal pressure was computed from equation (45), page 22, which reduces to

$$p = p_a + \delta\, h_1 +$$
$$\left[\tfrac{1}{2}\,\omega^2\,(\rho^2 - r^2_1) - (1 + \mu_1)\sin^2\gamma_1 + \left[\frac{k^2}{k^2_1}(1 + \mu_2) - 1\right]\sin^2\alpha\right]$$
$$x\,\frac{\dfrac{\delta}{g}\,\omega^2\,r^2_1}{\sin^2(\alpha + \gamma_1)}, \qquad (143)$$

with the following results for mid-section :

TABLE XII.

If	Then	Pressure lbs. per Square Foot.	Pressure lbs. per Square Inch.
$y = y_1$	$\frac{k_1}{k} = 1.205$	2,637	18.31
$y = 0.9\, y_1$	$\frac{k_1}{k} = 1.339$	2,630	18.26
$y = 0.8\, y_1$	$\frac{k_1}{k} = 1.510$	2,610	18.12
$y = 0.7\, y_1$	$\frac{k_1}{k} = 1.720$	2,581	17.92
$y = 0.6\, y_1$	$\frac{k_1}{k} = 2.01$	2,540	17.64
$y = 0.5\, y_1$	$\frac{k_1}{k} = 2.41$	2,380	16.52

In all cases where k is involved it is assumed that the stream completely fills the successive sections, otherwise k ought to be the actual section of the flowing stream :

But it is more important, in this case, to find the pressure at other points in the wheel. We have found the pressure, p_1, at the entrance into the bucket where $\rho = r_1$, and at $\frac{1}{4}$ the width of the crown from the inner rim, at $\frac{1}{2}$, $\frac{3}{4}$, and $\frac{7}{8}$, giving pounds per square foot as follows :

p_1	$p_\frac{1}{4}$	$p_\frac{1}{2}$	$p_\frac{3}{4}$	$p_\frac{7}{8}$	p_2
2648	2670	2667	2610	2420	2254

$$\frac{k_1}{k_\frac{1}{4}} = 0.986, \ \frac{k_1}{k_\frac{1}{2}} = 1.20, \ \frac{k_1}{k_\frac{3}{4}} = 1.67, \ \frac{k_1}{k_\frac{7}{8}} = 2.2, \ \frac{k_1}{k_2} = 3.3.$$

The pressure p_1 is at the gate, where the velocity is

$$V = 7.02 \div \sin 21° = 19.6 \text{ feet.}$$

At $\frac{1}{4}$ the width of the crown the velocity will be

$$v = 7.02 \times 0.986 = 6.29 \text{ feet,}$$

and the pressure is greater than at the gate. Since the terminal velocity is about 3.3 times the initial, the velocity ought to increase continually from the entrance into the bucket, to the exit, and therefore the normal sections should continually decrease, which, as is seen, may not be the case with flat crowns and partitions of uniform thickness. Thus, as shown just above, the initial section k_1 is 0.986 of the section at one-fourth the width of the rim from the initial element : hence the normal sections increase at first, and then decrease. To secure a continual diminution of sections, the depths of the buckets may at first diminish and later increase by curving the crowns, as in Fig. 21, the least depth being near the inner rim : or still more accurately as shown in Fig. 54a. If the wheel be cast the diminution of section may be secured by increasing the thickness of the walls, as in Fig. 22, the crowns being plane. If the terminal angle be less than 10 degrees, the outer depth (in this wheel) should be greater than the inner, and the crowns should flare outward, as in Fig. 23. The wheel may be so submerged as to produce any desired terminal pressure. If the Tremont wheel were submerged $394 \div 64.2 = 6.1$ feet more (p. 73), or 8.3 feet below the surface of the water in the wheel pit, then $p_2 = p_1$ or the pressure at the ends of the bucket would be equal.

47. *The direction of the water as it leaves the outer rim* is given by equation (35),

$$\cot \theta = \cot \gamma_2 - \frac{\omega\, r_2}{v_2 \sin \gamma_2}.$$

where θ is the angle measured from the rear arc. This gives, in this case,

$$\theta = 91°\ 33',$$

or the water will be thrown *forward* of the radius prolonged 1 degree 33 minutes; hence the direction should be nearly radial. Francis found, by a movable vane, about 34 degrees; but it is difficult to account for so large an angle, and one is inclined to think that there was a defect in the measurement. The measured angles appear to be very erratic. It is asserted by some writers that the quitting direction should be radial for best effect, but theory and experiment unite in showing that the assumption is not true, except in special cases.

According to the latter part of Article 45 the particles of a finite stream at quitting will not pass off in parallel lines, and may account, at least in part, for the discrepancy between the computed and measured angles.

48. *The radial* component of the velocity at quitting will be

$$v_2 \sin \gamma_2 = 5.990,$$

or, say, 6 feet per second. The velocity of the quitting water will be V_2 in equation (34), which is

$$V_2 \sin \theta = v_2 \sin \gamma_2;$$
$$\therefore\ V_2 = 5.99,$$

which is nearly the same as the radial component, as it ought to be, since the direction is nearly radial—call it 6 feet.

The initial velocity

$$v_1 = 7.1,$$

or, say, 7 feet, is radial in this case, so that the *radial* component of the velocity diminishes from about 7 feet to about 6 feet. For intermediate points it will depend upon the ring sections.

49. *To find the diameter of the shaft.*—The shaft will be subjected to a twisting stress. Let P be a twisting force, and a its arm in feet; n, the number of revolutions of the wheel per minute; $H.P.$ the number of horse-powers delivered, then

$$\frac{P.\ 2\ \pi a n}{33,000} = H.P.$$

$$\therefore 12\ Pa = \frac{63,000\ H.P.}{n} = \frac{1}{16}\ \pi\ J\ d^3,$$

where d is in inches, J the modulus of rupture to torsion $= \frac{1}{10}$ of 50,000, say.

$$\therefore d = \sqrt[3]{100\ \frac{H.P.}{n}}\ \text{nearly.} \tag{144}$$

If $H.P. = 161$, $n = 52$, then $d = 6\frac{3}{4}$ inches, nearly.

In the Tremont wheel, the diameter of the shaft was 7 inches from the wheel to the upper bearing, and larger in the hub. This gave a large margin for safety, the factor being more than 16.

50. *To find the path of the water in reference to the earth.*—Divide the space between the outer and inner rims into any number of parts, by concentric circles. Suppose there are 8 equal parts,

$$r_2 = 4.2$$
$$r_1 = 3.38$$
$$8)0.82$$
$$0.1025 \text{ feet between consecutive rings.}$$

Let ρ_1, ρ_2, etc., be the radii of the successive annuli, then

$$\rho_2 - \rho_1 = \rho_1 - r_1 = 0.1025 \text{ feet.}$$

The initial velocity being radial and $v_1 = 7.0$ feet, the time required to go from the inner circle to the next will be, considering the velocity uniform over this space,

$$T_1 = \frac{\rho_1 - r_1}{v_1} = \frac{.1025}{7.0} = 0.0146 \text{ seconds;}$$

but in this time the point a_1 of the bucket, Fig. 24, will have gone forward a distance

$$a_1\,d_1 = \rho_1\omega T_1 = 0.27 \text{ feet.}$$

The radial velocity at the second arc will be

$$v_{r_1} = \frac{2\pi r_1 - Nt \div \sin \gamma_1}{2\pi\rho_1 - Nt \div \sin \gamma} \cdot \frac{y_1}{y}\,v_1 = \frac{r_1\,y_1\,v_1}{\rho_1\,y} \text{ nearly ; (145)}$$

in which y_1 and y must be measured from the elevation in Fig. 19. It will be a little more accurate to take r_1 midway between r_1 and ρ_1, and y midway between ρ_1 and ρ_2; but as the working of the wheel and the efficiency do not depend upon the path of the water in reference to the earth, and only gives some information in regard to the course of the water, the method given is considered sufficiently accurate. Then

$$T_2 = \frac{\rho_2 - \rho_1}{r_{r_1}},$$

and

$$a_2\,d_2 = \rho_2\,\omega\,(T_1 + T_2).$$

and so on through the wheel. It will be seen from Fig. 24 that the terminal direction is nearly radial.

A still more simple, but approximate, method, is to consider the diminution of the radial velocity as uniform. The initial being 7 feet and the terminal 6 feet (say), the diminution of velocity from one annulus to the succeeding one will be $\frac{1}{8}$ of $(7 - 6) = 0.125$ feet. Then

$T_1 = 0.1025 \div 7$; $T_2 = 0.1025 \div 6.875$; $T_3 = 0.1025 \div 6.750$, etc.

(The Tremont turbine ran continuously from 1849 to 1892, when it was removed to give place for another wheel of larger power.)

51. *The Bucket.*—In the study of the Tremont turbine, something was said, in Article 45, in regard to the form of the bucket, but it is such an important element that more may properly be said. The parallel flow wheel offers the simplest solution. Let the hori-

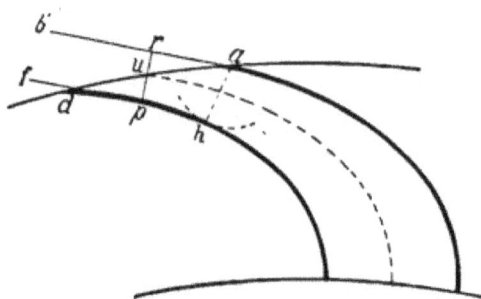

Fig. 20a.

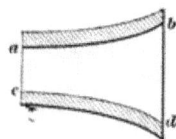

Fig. 23.

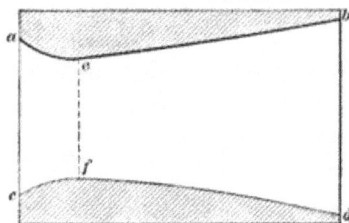

Fig. 21.

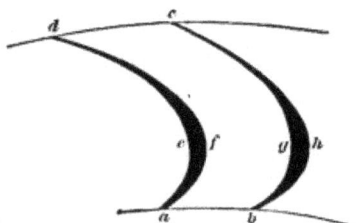

Fig. 22.

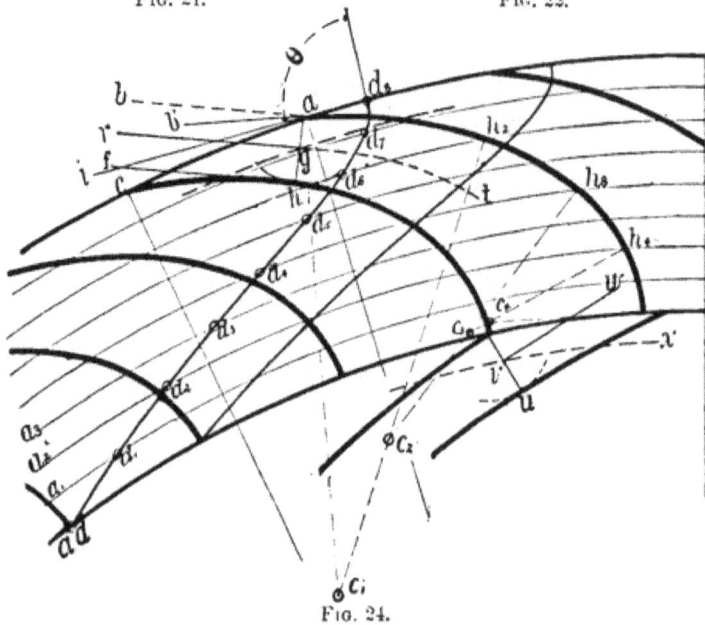

Fig. 24.

zontal lines through a and e, Fig. 25, represent respectively the lower and upper crowns of a parallel flow wheel, and the one through g the upper limit of the guides. The number of buckets being assumed and the diameter of the wheel given, the distances $a\,b = b\,c$, etc., will be found. At a, b, c, etc., lay off lines $a\,d$, etc., making angles $\gamma_2 = 15°$ or whatever value is assumed. From b erect the perpendicular $b\,d$, and prolong it to e; with e as a centre and radius $e\,d$ describe the arc $d\,f$, and similarly with the other buckets. This will make $\gamma_1 = 90°$ If it is desired to make γ_1 more than $90°$, the centre of the arc must be between e and d; and if less than $90°$, the centre must be on $e\,d$ prolonged. In a similar manner for the guides, lay off lines making angles $\alpha = 25°$ or whatever value is assigned; draw the perpendicular $f\,h$, prolong to g, and with g as centre and radius $g\,h$ describe an arc, and similarly for others. This is a simple mode of constructing the lines for the buckets and guides.

If γ_1 is obtuse, the normal section of the bucket, if the walls are of uniform thickness, will increase from the initial end to some point within the wheel and beyond that decrease. This, as has been stated, is an objectionable feature. This objection may be removed by making the back of the vane of suitable form, as in Fig. 26, and to make it light as possible it may be "cored." This gives the appearance of a double vane, and is called a "back vane," a feature first introduced by Haenel, a constructor of turbines, about the year 1848. A "back vane" considered by itself is objectionable on account of its weight and difficulty in construction. At the present time most turbines are made with an initial angle of the bucket of $90°$ or less, for which there is little need of back vanes, and they are now rarely made.

The surfaces of the vanes may be generated by a radius having as line directrices the axis of the shaft and one of the curves in Fig. 26, placed at the circumference, and the plane of the base of the wheel as a plane directrix. The surfaces will be warped —helicoidal.

When so generated the slope of the surface nearer the shaft will be greater than that more remote. By the use of back vanes this may be avoided, and almost any desired form of bucket made. The varying slope produces a difference in the mechanical action, and centrifugal force causes the outer part of the bucket to be full when the inner part may not be. For these and other reasons the buckets are sometimes divided into two or three smaller ones by annular rings. Fig. 27 is the plan of a bucket of uniform width.

Attempting a similar process for the radial outflow wheel, we proceed as follows:

Let $a\,b = b\,c = c\,d =$ width of bucket, Fig. 28. Through a, b, c draw lines, making angles equal to γ_2. From b erect perpendicular $b\,e$, and with o as a centre on $b\,e$ prolonged and $o\,e$ as radius describe the arc $e\,l'\,l\,i$, and similarly for the other buckets. Then from l draw $l\,p$ perpendicular to $o\,j$ as a partial back vane, and similarly for the others. The width of such a bucket is variable, increasing from $l\,j$ to $l'\,m'$, thence decreasing to f and again increasing to the end, which feature is objectionable. If the initial angle, γ_1, is to be 90°, $b\,e$ must be prolonged to meet $i\,O$ tangent at i, where the point i may be found by trial. The buckets may be made of uniform width as follows: Lay off the terminal angles, γ_2 at a, b, c, Fig. 29, as before. At e, a small arbitrary distance from b, erect the perpendicular $e\,o$, and describe $e\,f$. Similarly with o' as centre describe $h\,i$. From f erect $f\,l$ perpendicular to $o'\,h$, and with o' as centre describe $p\,l$ to the intersection with $o'\,i$, and from p draw $p\,l$ parallel to $c\,i$. The thicker part of the partition wall may be cored as indicated in the figure. Since the Boyden wheel with Russia sheet-iron plates for buckets, and the Swain and Hercules turbines, hereafter described, with cast-iron buckets, give high efficiencies and are durable, it is hardly necessary to give further discussion on forms.

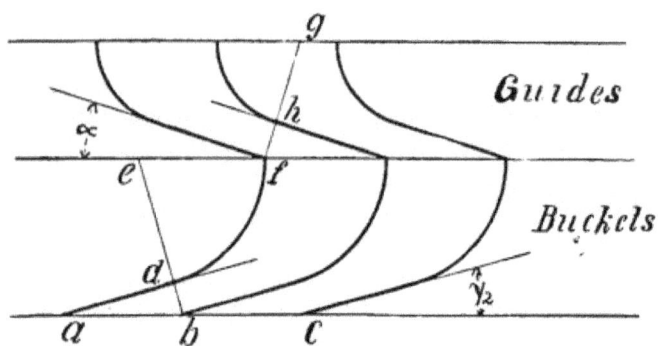

Guides

Buckels

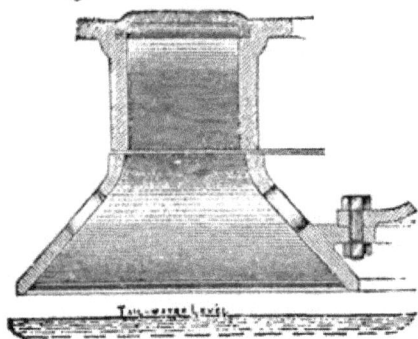

Fig. 25.

Fig. 26.

To find the depth of the buckets, when of uniform breadth, we have in all cases of continuity,

$$k\,v = k_1\,v_1 = k_2\,v_2,$$

and since the breadth, $i\,p$, Fig. 29, in this case is uniform,

$$y\,v = y_1\,v_1 = y_2\,v_2. \tag{146}$$

and if v increase by equal increments, y will decrease correspondingly, and the equation will be an equilateral hyperbola, in which the axis of v is the developed arc of the bucket. As shown in Tables VI., VII., IX., X., v_2 exceeds v_1; hence the terminal depth should be less than the initial for this case. This form of bucket has not, so far as known to the writer, been used, but it seems to have commendable features.

52. *The Boyden diffuser* shown in Fig. 30 consists of a conically diverging stationary piece, A, outside the buckets. Its office is to produce a diminishing velocity of discharge, and thus impart more energy of the water to the wheel.

53. *Turbine at Boott Cotton Mills*, Lowell, Mass.—This is an inward flow or vortex wheel, Fig. 31. It was made from the design of Mr. Francis in 1849, and tested by him in 1851, and was to develop 230 horse-power with 19 foot head.

The regulating gate was a cylinder of cast iron placed between the guides and wheel, and was made water-tight by means of leather packing. The water was conducted to the wheel by a wrought-iron riveted pipe, about 130 feet long, 8 feet in diameter, plates $\frac{3}{8}''$ thick.

DIMENSIONS OF THE BOOTT TURBINE.

Outside diameter,	$2r_1$	$= 9.338$ ft.
Inside "	$2r_2$	$= 7.987$ "
Least depth of guide passages,	Y	$= 0.999$ "
Outer " " buckets,	y_1	$= 1.000$ "
Inside " " buckets,	y_2	$= 1.230$ "
Number of buckets	N	$= 40$
" " guides,		40

Thickness of wheel vanes, ... $\tfrac{1}{4}$ in. $= 0.0208$ ft.

"   " guide   "   . . . . $\tfrac{3}{16}$ in. $= 0.0156$ "

Terminal angle of guides, ... $a = 8°$

"   "   " stream, mean, . . $a' = 12°$

"   "   " buckets, ... $\gamma_2 = 10°$

Mean angle of outflow from buckets, $\gamma'_2 = 15°$

Initial   "   " buckets, ... $a = 62°$

Measured area of guide passages, . . $K'' = 5.904$ sq. ft.

"   "   " outflow from buckets, $k'_2 = 6.8092$ " "

Calculated least area of buckets, . $k_2 = 4.343$

If contraction be 0.9 effective area, . $K'' = 5.3136$ sq. ft.

"   "   " 0.9 "   " $k_2'' = 6.1283$ " "

Ratio of depths, ... $\dfrac{y_1}{y_2} = 0.82$.

TABLE XIII.

ABSTRACT OF EXPERIMENTS ON THE BOOTT CENTRE-VENT WATER-WHEEL.

No. Exp.	Height of regulating gate in exp.	No. rev. per minute.	Total fall. *H*. Feet.	Quantity of water passing the wheel. Cu. ft. per second. *Q*.	Power of the water 62.421 *QH*. Ft. lbs.	Work by brake. Ft. lbs. *U*.	Efficiency. $\frac{U}{62.421\,QH}$	$\sqrt{gH}$	Velocity of exterior circumference. $= r_1$	Ratio of $\frac{r_1}{\sqrt{2gH}}$
5	3	30.0	14.197	67.020	59351.7	22630.5	0.38129		14.647	0.68470
6	3	27.0	14.143	66.889	59002.2	22026.8	0.37332		13.185	0.43714
14	6	41.7	13.778	90.166	77481.4	45135.3	0.58254		20.372	0.6 433
15	6	35.9	13.606	91.697	77812.1	46402.8	0.59634		17.540	0.50291
23	9	43.4	13.332	103.769	86229.3	62065.4	0.71986		29.861	0.71292
24	9	42.7	13.334	103.689	86229.3	62752.2	0.72774		29.407	0.70067
25	9	41.9	13.304	104.229	86483.7	63199.3	0.73007		36.785	1.28907
27	12	42.6	13.400	112.525	94037.5	74979.3	**0.79716**		20.806	0.70868
28	12	42.0	13.431	112.087	**94662.2**	75347.2	0.79596		20.515	0.69796
29	12	40.7	13.331	112.562	93603.9	74580.9	0.79707		19.929	0.68058
30	12	40.3	13.387	112.996	94206.4	75158.2	0.79699		19.711	0.67139
31	12	39.6	13.386	113.071	94415.2	75208.3	0.79657		19.364	0.65990
32	12	38.9	13.383	113.164	94471.1	**75249.2**	0.79658		19.029	0.64856
33	12	38.2	13.356	113.050	94219.0	75142	**0.79753**		18.688	**0.63692**
34	12	37.4	13.381	113.673	94881.9	75103	0.79754		18.316	0.62431
35	12	36.8	**13.405**	114.293	95571.2	75202	0.78132		17.998	0.61291

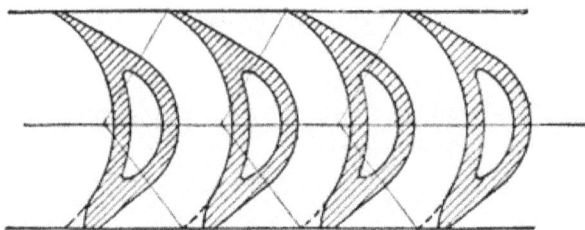

FIG. 27

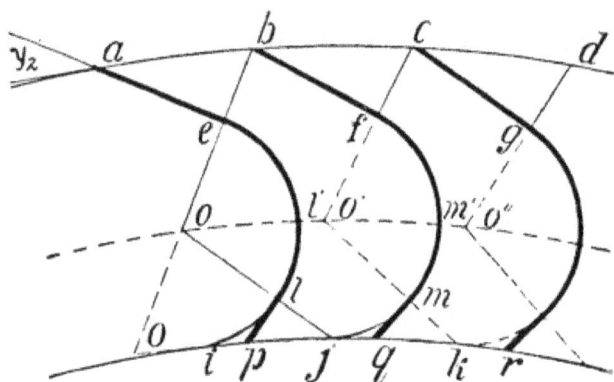

FIG. 28.

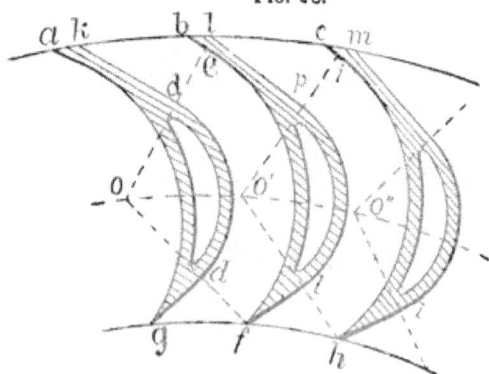

FIG. 29.

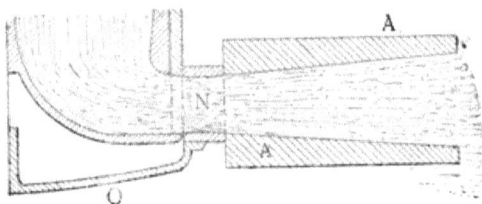

FIG. 30

54. *Computations.*—An examination of Fig. 31 shows that a tangent to a guide at its terminus falls entirely outside the wheel; hence the angle α, which should be used in the formula, is indeterminate. If the water were not confined none would enter the wheel, but as it is confined and the supply continuous, it will be forced into the wheel, but at unknown angles some filaments may enter at 90°. It is really a case of discontinuity as regards our formulas, and hence they do not strictly apply; knowing, as we do, the least section between the guides, 5.904 square feet, and the volume discharged, 113 cu. ft. per second, the velocity of exit from the guides will be

$$V = 113 \div 5.904 = 19.14 \text{ feet per sec.}$$

The velocity of the wheel for best effect, experiment 33, was

$$\omega' \, r_1 = 18.688 \text{ feet,}$$

and since $\gamma_1 = 62°$, equation (17) will give $\alpha = 16°$ about, which is somewhat larger than the mean angle given above. Then

$$v_1 = \frac{V \sin \alpha}{\sin \gamma_1} = \frac{19.14 \times 0.27}{0.866} = 6 \text{ feet nearly.}$$

But this is an inverse process, and we now try a direct process, assigning to μ_1 and μ_2 large values on account of the resistances resulting from imperfect conditions.

TABLE XIV.

DATA.

$$r_1 = 4.67, \ \tau_2 = 4.00, \ \gamma_2 = 15°,$$
$$H = 13.356 \text{ feet}, \ Q = 113.09 \text{ cu. feet.}$$

REQUIRED.	$\alpha = 15°$ $\mu_1 = 0.20$ $\mu_2 = 0.20$ $\gamma_1 = 60°$	$\alpha = 12°$ $\mu_1 = 0.25$ $\mu_2 = 0.20$ $\gamma_1 = 62°$	Measured Values.
Angular velocity, Eq. (16), ω	4.442	4.288	3.996
Rev. per minute	42.42	40.95	38.16
Efficiency, Eq. (16a), E, per cent	82.50	81.06	79.753
Initial velocity (18), v_1	5.557	4.33	.....
Terminal velocity (19), v_2	17.339	17.03	.....
Inner depth (141), y_1	0.828	1.062	1.000
Outer depth (141), y_2	1.150	1.181	1.230
y_2/y_1	1.37	1.112	1.230

A " Boyden " Turbine.

55. In 1882 a Boyden turbine of the Fourneyron type was tested for the Merrick Thread Company. This was several years after the construction of the most celebrated Boyden wheel. The following are the dimensions of the wheel, and results of the test :

Internal diameter, ... $2r_1 = 73.60$ in.

External " . . . . . . $2r_2 = 90.00$ "

Terminal vane angle, ... $\alpha = 24°$

Mean exit angle, ... $\alpha' = 29°$

Initial angle of bucket, ... $\gamma_1 = 90°$

Terminal angle of bucket, . . . $\gamma_2 = 26°$

Mean angle of flow, ... $\gamma'_2 = 29°$

Initial depth of wheel, . . . $y_1 = 8.64$ in.

Terminal depth of wheel, . . . $y_2 = 9.125$ in.

Number of buckets, . . . $N = 34.$

" " guides, ... 54.

Terminal least section of guides . . $K'' = 6.814$ sq. ft.

" " " " buckets . . $k'_2 = 5.660$ " "

TABLE XV.

ABSTRACT OF A TABLE OF FRANCIS EXPERIMENTS WITH A BOYDEN WHEEL (OF FOURNEYRON TYPE).

Gate Open-ing.	Number of rev. per minute.	Head, Feet. H.	Vol. water per second, Cu. ft. Q.	Horse-power measured by Brake, Ft. lbs. $H.P.$	Efficiency per cent. K.	Ratio, $\frac{2\pi r_1 n}{\sqrt{2gH}}$
1.000	57.00	16.61	145.35	216.79	79.17	0.560
1.000	59.33	16.57	146.18	216.55	78.82	0.584
1.000	63.50	16.60	147.10	222.04	**80.17**	**0.624**
..000	66.50	16.62	148.32	220.28	78.79	0.653
0.833	55.67	16.66	142.04	208.19	73.71	0.546
	58.75	16.63	142.82	207.28	76.93	0.577
0.773	56.00	16.74	136.12	188.94	73.11	0.548
0.662	55.75	16.80	128.86	169.28	72.42	0.545
0.442	50.63	17.10	105.65	113.36	55.32	0.461

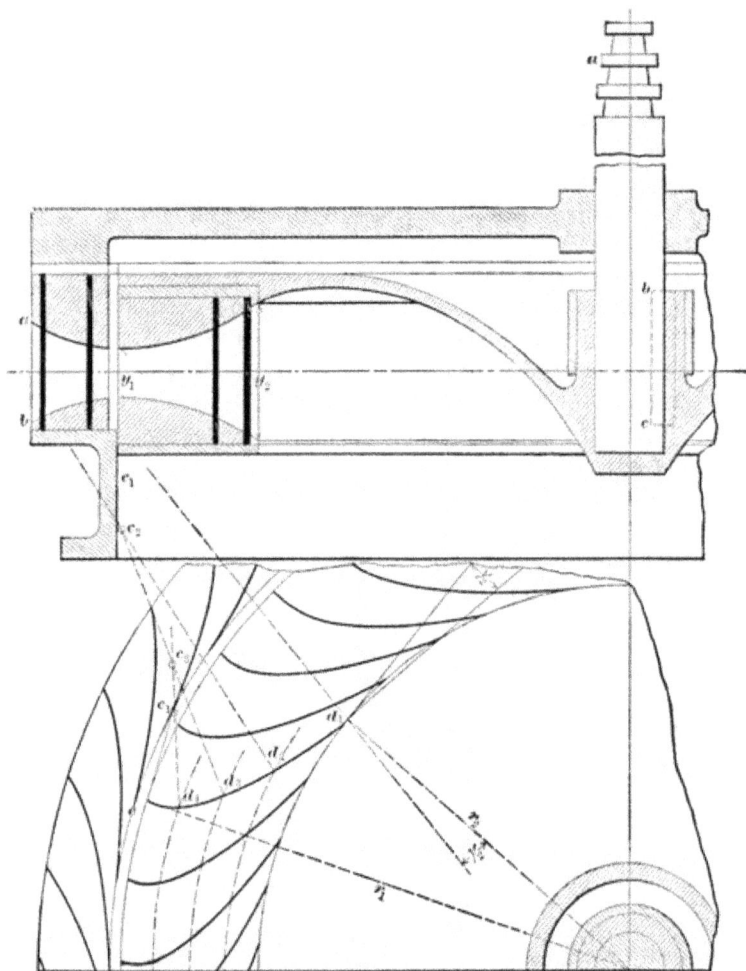

Fig. 31.

The hydraulic efficiency would probably be about 82 per cent. Judging from the analysis of the Francis wheel and the comparatively large terminal vane angles in this wheel, the prejudicial resistances were low ; probably $\mu_1 = \mu_2 = 0.10$ about.

56. It will be seen from the preceding tests that the efficiency falls off quite rapidly as the gate is closed more and more. This is a peculiarity of the pressure turbine, and for this reason is objectionable where the supply of water is not sufficient to fill the buckets with open gate, for the lack of such a supply necessitates a partially closed gate to secure a head in the supply chamber. This condition of things induced Fourneyron to divide the wheel into several chambers, as if separate wheels were placed one above the other, as in Fig. 32. With this arrangement, with the gate one third opened, the lower third would work as if it were the only wheel ; similarly the second third, and finally, when the gate was completely open, the three parts worked as one wheel.

57. *Fourneyron Turbine at St. Blaise.*—One of the first turbines constructed by Fourneyron was erected at St. Blaise, and worked under the high head of 354 feet, developing 30 effective horse-power at 2300 revolutions per minute.

Outer diameter $r_2 = 12.99$ inches.
Inner " $r_1 = 7.47$ "

Later two larger wheels were substituted for this one working under the same head, developing 60 horse-power with the same number of revolutions. The outer diameter was $r_2 = 21.66$ inches. At these high speeds the bearings needed renewing every 10 to 14 days. They discharged above the water, as shown in Fig. 33. Later these were replaced by tangent wheels, and still later by Girard turbines.

58. *Parallel Flow Turbines.*—The parallel flow wheels are usually classed as Jonval or Girard, which are constructed sub-

stantially alike ; but the former is buried or submerged, while the latter discharges above the water in the wheel pit, Fig. 34. The French engineers Callon and Girard, about 1856, began to design impulse turbines for all possible conditions—high and low falls, large and small quantities of water, axial and radial flow, with horizontal, vertical, and inclined axes. All turbines in which the velocity from the supply chamber into the wheel was that due to the head, which we have called wheels of *free deviation*, were called *impulse turbines*, and in Europe every variety of impulse turbine frequently goes by the name of "Girard." He was the first to ventilate the buckets, so that the pressure in the wheel would be that of the atmosphere, Fig. 35.

A A, Fig. 34, are buckets placed between parallel crowns and supported by radial arms or segments attached to the shaft ; *B B* are guide vanes, directrices, or distributors. The cross-section of Girard's turbines are bell-mouthed ; but as first made by Jonval, the sides were parallel, as shown in Fig. 36. The Jonval turbine may be placed at any point between the level of the water up stream and that in the tail race, provided there be a closed tube, called a "suction tube," through which the water passes before being discharged, as in Fig. 36.

59. *The "Collins" Turbine.*—The Collins turbine is the only parallel flow wheel of American make for which we have the results of a test. It was tested in 1883.

DIMENSIONS.

$$\alpha = 17\tfrac{3}{4}°, \; \gamma_1 = 77\tfrac{3}{4}°, \; \gamma_2 = 19°.$$

Mean diameter, ...	$2r_m$	$= 4.170$ ft.
Terminal depth of guides, ...	Y	$= 0.836$ "
Number of buckets, ...	N	$= 24$
" " guides, ...	N_1	$= 30$
Measured terminal arc of guides, .	K'	$= 2.912$ sq. ft.
" " " " buckets, .	k_2	$= 2.882$ " "

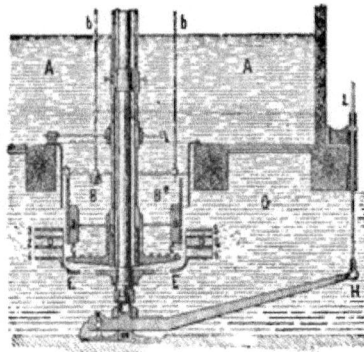

Fig. 32.

Fig. 33.

TABLE XVI.

ABSTRACT OF TEST OF A 60-INCH COLLINS TURBINE SUBMERGED IN THE WHEEL PIT (JONVAL TYPE).

GATE OPEN-ING.	Total head. H.	Volume of water per second. Q.	Horse-power measured by brake. $H.P.$	Efficiency. Per cent. E.	Revolutions per minute. N.	Ratio. $\frac{N}{\sqrt{2gH}}$
1.000	16.57	64.35	96.41	79.85	85.25	0.573
1.000	16.55	64.40	97.03	**80.40**	89.17	**0.600**
1.000	16.56	64.40	96.65	80.03	93.00	0.625
0.748	.79	58.50	86.69	77.49	81.50	0.544
0.600	.86	53.61	69.53	67.93	77.90	0.650
0.503	17.18	45.14	56.22	64.02	69.67	0.460
0.303	17.41	34.25	34.72	51.42	71.00	0.218
0.161	17.84	24.80	19.12	38.16	53.00	0.984

The same turbine was tested with a suction tube, with the following results:

TABLE XVII.

SIXTY-INCH COLLINS TURBINE WITH A SUCTION TUBE.

GATE OPEN-ING.	Head feet. H.	Volume of water cu. ft. per second. Q.	Brake power. ft. lbs. $H.P.$	Efficiency. Per cent. E.	Ratio. $\frac{N}{\sqrt{2gH}}$
1.000	16.56	64.88	102.18	84.01	0.604
1.000	16.55	64.99	102.70	**84.34**	**0.646**
1.000	16.56	64.88	101.25	83.25	0.709
0.548	17.01	68.87	64.83	65.97	0.457

The same turbine was also tested with a suction tube, in which was a cone with its base placed just under the centre of the wheel, so as to act as a diffuser, with the following results:

TABLE XVIII.

SIXTY-INCH COLLINS TURBINE WITH SUCTION TUBE AND INVERTED CONE.

Gate opening.	Head, feet. H.	Vol. water cu. ft. per sec. Q.	Brake power. $H.P.$	Efficiency. Per cent. E.	Ratio. w/s $\sqrt{2gH}$
1.000	16.56	65.18	101.46	83.05	0.578
1.000	16.57	65.30	102.62	**83.76**	**0.606**
1.000	16.58	65.37	101.73	82.92	0.712
0.548	16.96	52.09	69.66	69.66	0.492

The efficiency was increased with the suction tube, but not with inverted cone.

60. A 96-inch "Collins" (parallel flow) turbine was tested at the Holyoke testing station about 1883, which gave the highest efficiency of any recorded experiments with a parallel flow wheel so far as we know.

NINETY-SIX INCH COLLINS TURBINE (JONVAL TYPE).

Gate opening, ... fully open.
Total head, ... $H = 16.59$ ft.
Volume of water per second. . . . $Q = 113.46$ cu. ft.
Number of revolutions per minute, . $N = 63.38$
Brake horse-power, ... $HP = 131.49$
Efficiency of wheel, per cent., . . $E = 85.06$
Hydraulic efficiency, possibly, . . 87 per cent.

61. *Segmental Feed.*—A turbine works with better effect when the buckets are full; and when the wheel is too large to secure, at all times, full buckets with variable supply, means have been devised to shut off a part of them, leaving others fully open. Fig. 37 shows such a device, where the upper part of the guide forms the gate by sliding downward to close the passage of the guides, or distributors. Those at the right marked B are fully open, while those marked A at the left are shown partly closed. They may be so constructed that two or three gates may be closed at a

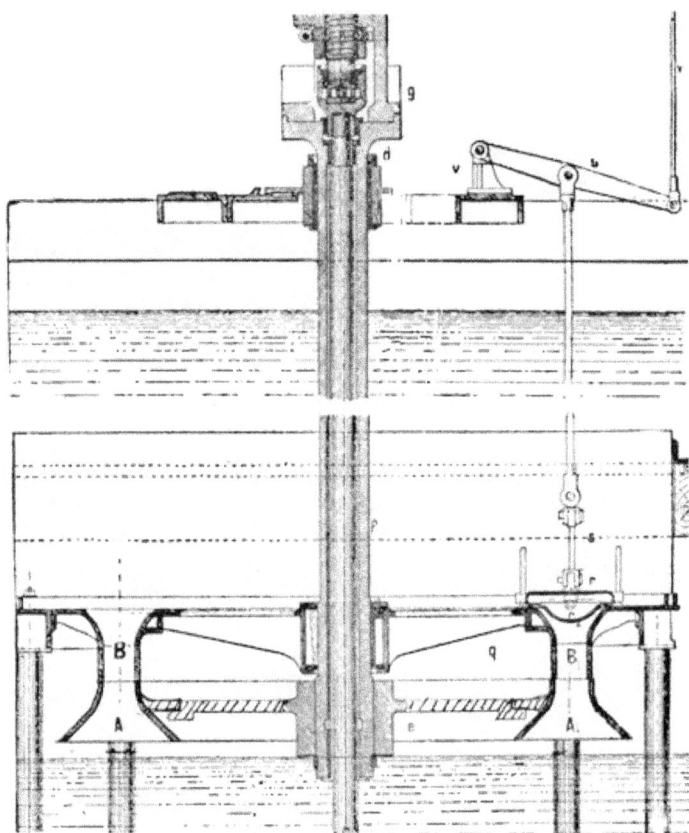

Fig. 34.

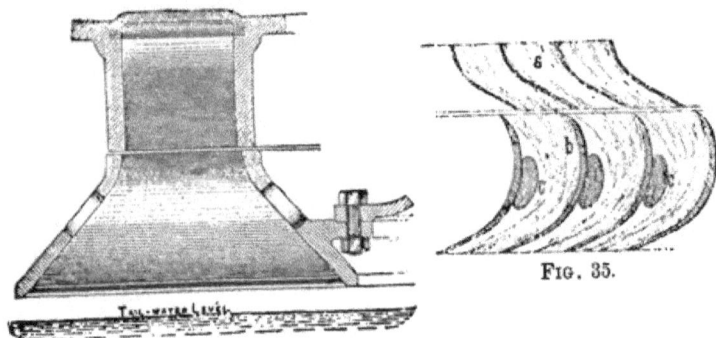

TAIL-WATER LEVEL.

Fig. 35.

Fig. 36.

time, while the others remain open, and so be adapted to great range of the supply of water. Segmental feed does not produce quite as efficient a wheel as full feed, but, as will be seen, the difference is not as great as might be anticipated, considering that a portion of the water will have but little effect during the filling and emptying of the buckets.

62. *Haenel's Turbine.*—Haenel, manager of machine works at Magdeburg, constructed several parallel flow wheels, and made an extensive set of experiments which are very instructive. Admission to the guide wheel was regulated by a pair of rubber strips, supported by iron stays, and rolled upon two conical rollers, the object being to admit water to as few passages as desired, or to open all 32 of them at the same time. Fig. 38. *A A* are the buckets, *B B* the guides, *C C* conical rollers, *E E* the rubber flap. The wheel passages were designed to be of equal sections normal to the flow of the water, so that the velocity through the wheel would be uniform when submerged. This was accomplished, in part, by the use of back vanes. The ventilating pipes connected with the vane passages were sometimes open and sometimes closed, but there was no appreciable difference in the efficiency due to the difference of the two conditions.

The turbine tested was one of eight, all alike, having the following dimensions. The dimensions in feet are approximate, as the original were in Prussian feet, and for our purpose it is not considered necessary to be particular about the fractions of an inch in English units :

1 Prussian foot = 1.02972 English feet.
1 square Prussian foot = 1.06032 English square feet.
1 cubic Prussian foot = 1.09183 English cubic feet.

DIMENSIONS OF HAENEL'S TURBINE.

Outer diameter of wheel and guide at inflow,				5 ft.	9¼ in.
Inner " " " "				4 "	6 "
Mean " " " "		$d = 5$ "	1½ "

Outer diameter of wheel at outflow, . . .   6 ft. 5 in.

Inner   "   "   "   . . . .   3 " 10 "

Initial depth (width) of wheel buckets,   . . $y_1 = 1$ " $3\frac{1}{2}$ "

Terminal   "   "   "   . . $y_2 = 2$ " 7 "

Depth of wheel, ...   1 " $\frac{1}{2}$ "

Angle of outflow from guides, ... $\alpha = 22° 30'$

Initial angle of buckets, ... $\gamma_1 = 45°$

Terminal angle of buckets on concave side, . $\gamma_2 = 26° 20'$

     "   "   "   " convex " . $23° 00'$

Measured area between guides, ... $K' = 3.47$ sq. ft.

Effective area 0.9 of measured. . .   . $K'' = 3.12$ " "

   "   " of buckets at outflow, . $k_2'' = 6.38$ " "

TABLE XIX.

RESULTS OF TESTS OF HAENEL'S TURBINE.

Number of gate openings.	Head, feet, English. H.	Depth immersed. Feet.	Volume of water discharged per second, English cu. ft.	Energy of fall. Prussian ft. lbs. $\delta Q H$	Brake power per sec. Prussian. L	Revolutions per minute. n.	Brake efficiency. E.	Hydraulic Efficiency.
4	6.65	0.00	5.80	2250.8	1457.7	29.0	0.6476	0.7101
8	6.58	0.11	12.98	5052.5	3340.6	27.5	0.6612	0.6876
12	6.48	0.21	19.80	7540.9	5351.2	36.5	0.7096	0.7331
16	6.44	0.21	26.10	10217.0	6369.9	35.0	0.6816	0.6982
24	6.18	0.39	47.61	15844.0	10556.0	31.5	0.6662	0.6759
32	6.26	0.29	53.66	19765.0	1214.8	72.5	0.0615	0.0703
32	5.90	0.54	62.20	21579.0	14703.0	39.0	**0.6813**	**0.6901**
			Another test.					
32	4.07	2.03	48.93	12060.	8042.5	32.0	0.6669	
			Still another test.					
32	5.12	1.48	45.66	14156	9676.2	33.0	**0.6836**	**0.6949**

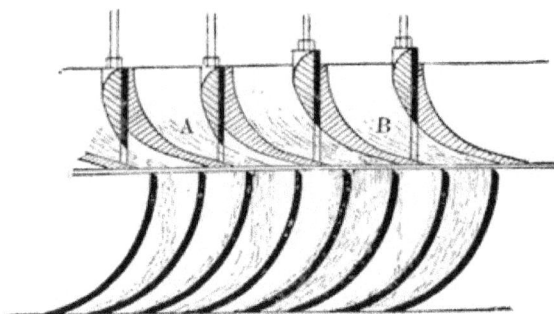

FIG. 37.

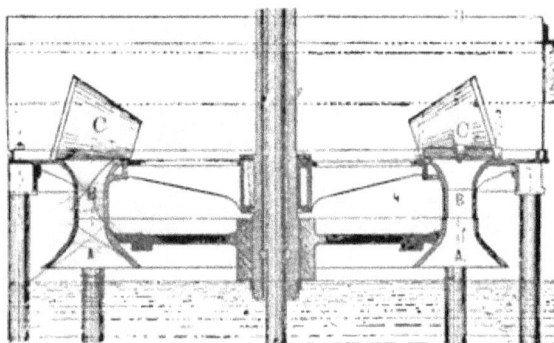

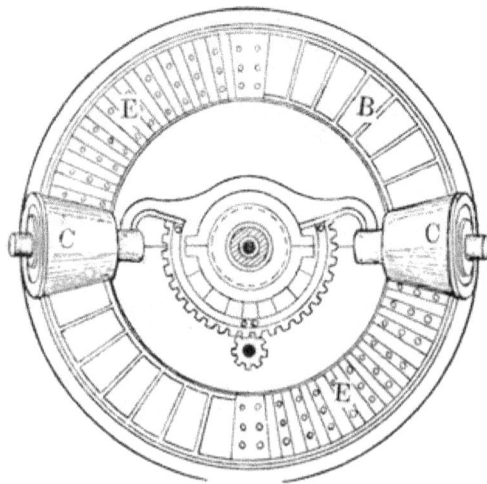

FIG. 38.

According to these results, the prejudicial resistances were $6\frac{1}{4}$ per cent. with 4 buckets working, and was about 1 per cent. with all the gates open. One test gave only $\frac{3}{4}$ of one per cent. prejudicial resistance.

63. *Tangential wheels* are radial with segmental feed. They are more especially adapted to high heads with a limited supply of water. They are made larger in diameter than pressure wheels of the same capacity, and hence, when the speed of the periphery is the same, the revolutions will be less in the same time. They may be inflow, as in Fig. 39, or outflow, as in Fig. 40.

64. *Industries* gives an interesting account of the turbines used in the steel-manufacturing establishment at Terni, Italy. They are segmental feed, radially outward flow, of free deviation, Fig. 40.

TABLE XX.

SOME DIMENSIONS OF TURBINES AT TERNI.

Head. H.	Horse-power. H.P.	Volume of water. Q.	Rev. per minute. n.	Inner diameter. r_1.	
Ft.		Cu. ft. per sec.		Ft.	In.
595.5	1000	19.77	210	7	10.5
595.5	800	15.89	200	8	2.4
595.5	500	9.89	240	6	5.9
595.5	350	7.06	200	7	10.5
595.5	150	3.00	250	6	4.7
595.5	50	0.99	850	1	10.25
595.5	50	0.99	850	1	10.25
595.5	40	0.85	450	3	6.1
595.5	40	0.85	450	3	6.1
595.5	30	0.60	600	2	7.5
595.5	20	0.42	450	3	6.1

65. Dimensions of a segmental feed, tangential turbine, radially outward flow of free deviation, with horizontal axis, Fig. 40.

Inner diameter, ... $2r_1 = 7$ ft. $10\frac{1}{2}$ in.
Outer " . . . . . . $2r_2 = 8$ " 11.1 "
Width of guide passages, . . . $Y = 4.32$ in.
" buckets, initial, . . . $y_1 = 4.72$ "
" " terminal, . . $y_2 = 15.75$ "
Number of wheel vanes, $N = 110$
Thickness of steel vanes, $t = 0.2$ in.
Length of supply pipe, ... $4642\frac{1}{2}$ ft.
(3481 feet of cast iron, remainder wrought iron.)
Thickness of cast-iron pipe, . 0.71 in.
Diameter " " . . 21.66 "
" of wrought-iron pipe, . 18.9 "
Thickness " " " . 0.20 " to 0.47 in.
Designed for, ... 400 H.P.
With head of, ... $H = 570.8$ ft.
And water supply of, ... $Q = 8.5$ cu. ft. per sec.

TABLE XXI.

TEST OF THE PRECEDING WHEEL AT IMMERSTADT. (AT THE TIME OF THE TEST 8.5 CU. FT. WERE NOT AVAILABLE.)

GATE OPEN-ING, INCHES.	No. of rev. per minute, n.	Head, feet, H.	Volume of water cu. ft. per second, Q.	Total power of the water, $\frac{\delta QH}{550}$	Brake power, $H.P.$	Brake effi-ciency, k_2
10	211	570.84	2.111	136.8	81.5	0.595
15	211	570.84	3.350	217.1	144.3	0.665
20	214	570.84	4.375	283.4	196.8	0.694
25	216	570.84	5.187	336.0	253.9	0.755
31	210	570.84	6.840	440.9	336.8	**0.764**

The power absorbed by the friction of the shaft was 5 H.P. with 10-inch opening of gate and 4.5 with 25-inch opening, or about 3 per cent. of the total power in the former case and

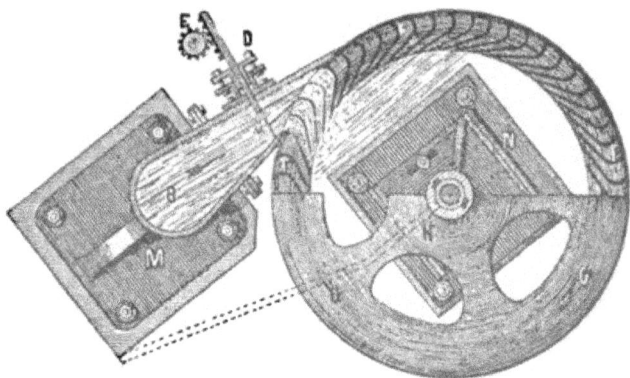

FIG. 39.

FIG. 40.

1 per cent. in the latter; or about 5½ per cent. of the brake power to less than 1½ per cent.

Turbines acting under a fall of about 650 feet are described by Knoke. They are radially outflow and of cast iron. They receive water through two guides at opposite extremities of the diameter (four guide passages in all).

DIMENSIONS OF TWO OF THEM.

Inner diameter, ...	11.81 in.	11.81 in.
Outer "	16.41 "	14.84 "
No. of vanes, ...	45 cast iron	84 wrt. iron.
No. rev., ...	583.	928
Vol. water per min., ...	4.52 cu. ft.	5.085 cu. ft.
Head, feet, ...	129.	131
Power of the water, $\delta \, Q \, H$, ft. lbs.,	36440.	53806
Brake power, ...	16498.	28809
Efficiency, ...	45.27.	53.5

MIXED FLOW.

66. *Risdon Wheel.*—Mixed flow wheels, so far as known, are inward and downward, of which the "Risdon," Fig. 41, is a type, giving also a view of the cross-section of the wheel just under the upper crown, in which $B \, D$ are guides and A buckets. The illustration shows the construction so clearly that a detailed description seems unnecessary. This gave the highest efficiency of any wheel tested at the Centennial Exposition, 1876, it being reported as 87 per cent. Those tests lasted but a few minutes, and may have been fortunate in showing a higher efficiency than they would maintain for a long period; but this wheel has maintained a reputation for high efficiency, and in some more recent tests has been reported as giving even 90 per cent. But all such high figures should be received with caution, and writers discard the last figure as being too improbable to be true.

67. *The Humphrey Wheel.*—This is an inward and downward
flow wheel, Fig. 42. The wheel *A A* has 13 buckets, *a a* extending below the casing. *C C* is a regulator, also containing 12
guides, the whole moving on spherical rollers, and operated by a
hand wheel, *G*, through shafts and bevel gearing *F*. By this
means water is admitted to or cut off from the wheel. The
wheel is enclosed in metal having curved surfaces, into which the
water was conducted by a riveted pipe. The lower end of the
shaft had a pivoted step, while the upper end was supported by
collar bearings. Fig. 43 shows a plan of the wheel casing, con-
duit, and brake used in the test. The diameter of the brake
pulley was 5.44 feet; width, 2.7 feet; arm of brake lever, 15.9
feet. The wheel was a design of the "Humphrey" Machine
Co., of New Hampshire, and was designed to deliver 270 horse-
power with a fall of 13 feet. It was tested by Mr. James B.
Francis about 1880. The head was determined by the difference
in heights of the water in two tubes, one entering the supply shaft
above the wheel and the other in the tail race; and the quantity
of water by means of weirs and hook gauges. There being no
contraction at the ends of the weir, the quantity was computed
from the formula

$$Q = 3.33 \, L \, H^{\frac{3}{2}},$$

where

    L = length of weir in feet = 11.92 for one weir, and 10.98
        for the other.

    H = head of water above sill in feet, measured some dis-
        tance above weir.

DIMENSIONS.

Outer diameter, $2r_2 = 8.1$ feet; inner, $r_1 = 2.0$; $\alpha = 15°$,
$\gamma_1 = 80°$. γ_2, not given, but the result of the test indicates
that it was small, probably about 12°. $Y = 2$ feet.

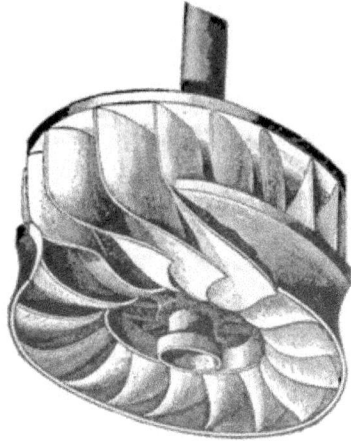

FIG. 41.

TABLE XXII.
TEST OF HUMPHREY TURBINE.
Brief extract of results.

OPENING OF REGULATOR.	Head. H.	Volume Water per second. Q	Total Water Power. $6\,QH$.	Brake Power. L.	Efficiency. K.	Ratio. $\frac{v_2}{v} \frac{r_2}{\sqrt{2\,gH}}$
Per cent.	ft.	cu. ft.	Ft. lbs.	Ft. lbs.	Per cent.	
101.00	12.478	207.82	161744	132475	81.9	0.7301
88.50	12.839	181.59	145421	111527	76.69	0.7831
68.21	13.103	166.80	136326	99084	72.68	0.687
52.90	13.456	131.52	110387	67363	61.02	0.7508
40.66	13.688	110.70	94520	53065	56.14	0.6702
27.94	14.069	79.26	69558	18517	26.62	0.8073
19.15	13.998	59.02	51534	8264	16.04	0.7225

The velocity—about 3.458 feet—in the supply pipe was neglected in determining the head, for which, if a correction be applied, it will reduce the brake efficiency to about 80.59, and hence the hydraulic efficiency would be about 82 per cent.

68. *The Swain Turbine.*—The Swain turbine is a mixed flow wheel, inward and downward, of which a sectional view is shown in Fig. 45, and an external view with vertical shaft in Fig. 44. W, Fig. 45, is the wheel, A the guides which are secured to the gate G, and are raised and lowered with it, and pass into and out of the chamber E. By this arrangement the inner ends of the guides are brought nearer the buckets than when the gate is between the bucket and guides, the space in this wheel being 1⅜ inches. The gate is opened by being lowered, so that the water first enters just under the crown, and passing inward and downward is discharged nearly radially, while that which enters lower is discharged nearly axially. The guides and upper ends of the buckets are shown in Fig. 46. There are three heavy cast-iron guides, one of which is shown in Fig. 46, through which pass the lifting rods, as shown at *e*. The other 21 guides were 0.23 inch thick of bronze, sharpened at the ends to 0.04 of an inch thick, nearly 19 inches long. It had 25 buckets of bronze, pressed into shape and cast

into a crown above and band below, and were 23.35 inches deep.
The gate G was made of two cylinders M and N, joined by a
disc, Q. At the lower end of M is a narrow flange, to which was
attached a leather packing to prevent the escape of water. The
leakage was so small as to be unimportant. The wheel rested on
an oak pivot, S, conical at its ends, free to turn or rest, and supplied with water through the pipe f. It was so arranged between the connecting piece a and crown V, that the step S could
be replaced by another, and the adjustment of the height of the
wheel made by means of the screws t, t. Opposite the thick
vanes are stationary supports, of which one is shown at O, Fig.
45, resting on the cast-iron base C, and support the chamber E
and cover of the wheel L. The openings in the base C for the
passage of water into the wheel pit flow outward, as shown in
the right-hand part of the figure.

An elaborate report of the test of this turbine—designed to
replace the centre-vent turbine established in 1849 at the Boott
Cotton Mills, previously described—is given by Francis and
published in the *Journal of the Franklin Institute* for April,
1875, to which we are, through the courtesy of the Swain Manu-
facturing Co., indebted for the facts and tables herein given.
Table XXIII. contains only the best result for each gate opening
up to "full gate." Mr. Francis states that Table XXIV. is
made from a curve of the results plotted on section paper.

It will be seen that the highest efficiency was obtained with
the gate closed, 1.08 inches. This indicates that there may have
been interferences of the stream of water in the bucket at full
gate opening. It is also worthy of special note that the efficiencies are well maintained down to the smallest opening of
2 inches. The hydraulic efficiency for the best test was about
85 or 86 per cent.

DIMENSIONS.

Outer diameter, ... $2r_1 = 6$ ft.
Least inner diameter, ... $2r_2 = 2\frac{2}{3}$ ft.

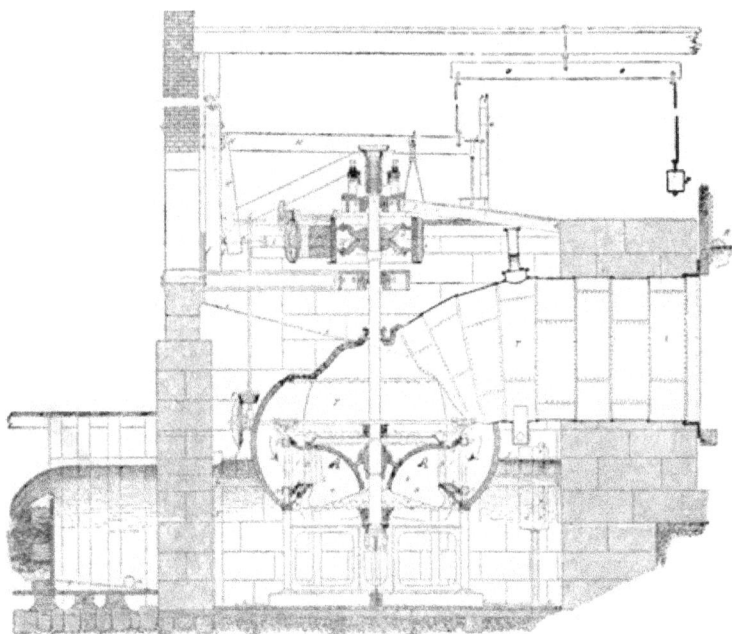

FIG. 42.

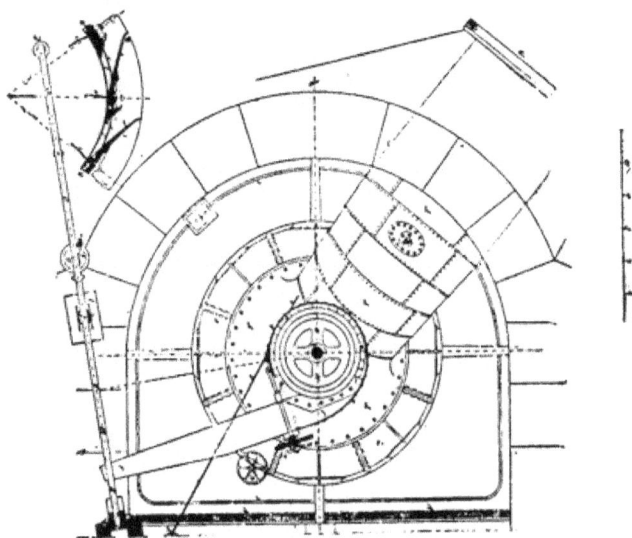

FIG. 43.

Depth of guide passages, . . . $Y = 13.08$ in.
"   buckets at inflow, . . $y_1 = 13.285$ in.
Mean terminal angle of guides, . $\alpha = 25°$
"   initial angle of buckets, . . $\gamma_1 = 90°$
"   terminal angle of buckets,
     estimated, ... $\gamma_2 = 25°$
Number of buckets, ... $N = 25$
"   guides, . . . $N_1 = 24$
Thickness of guides, . . $\begin{cases} t = 0.23 \text{ in. to} \\ t = 0.04 \text{ in. at end.} \end{cases}$
Measured area between guides, . $K' = 9.880$ sq. ft.
Least *measured* area of the
    buckets, ... $k_2' = 9.558$ " "

Where the water leaves the wheel axially (outer diameter), $\gamma_2 = 26°$, and where it leaves it radially, $\gamma_2 = 22°$, so it is estimated that the effective angle is about $25°$.

TABLE XXIII.

ABSTRACT OF THE RESULTS OF EXPERIMENTS WITH "SWAIN" TURBINE WHEN RUN FOR BEST EFFECT.

OPENING OF GATE. In.	No. of rev. per minute. n	Head, Feet. H	Vel. Water passing weir per second. Cu. ft. Q	Energy of the fall Ft. lbs. $8 Q H$	Brake Power Ft. lbs. U	Efficiency per cent. E	Ratio. $\frac{r_2 \cdot \omega}{\sqrt{2 g H}}$
2	60.5	14.283	51.177	45518	21572.5	47.39	0.6274
3	60.3	13.977	68.000	59189	34806.5	58.81	0.6623
4	60.0	13.609	83.804	71562	47072.5	66.02	0.6478
5	60.12	13.482	97.790	82048	59179.7	72.13	0.6537
6	60.21	13.281	110.002	91044	62263.5	76.08	0.6680
7	66.33	13.102	119.268	97309	76962.3	79.06	0.7184
8	60.40	12.968	130.253	105184	84794.3	80.61	0.6979
9	60.5	12.836	138.839	110056	91961.2	82.88	0.7008
10	60.43	12.680	144.774	134202	95588.9	83.63	0.7082
11	60.56	12.640	151.501	130226	99206.1	83.26	0.7227
12	67.6	12.581	156.703	122158	102435.2	**83.85**	**0.7490**
13.08	108.36	13.099	130.392	98204	30237.7	20.61	1.1738
Full Gate.	97.8	12.880	137.871	110580	53189.4	49.30	1.0682
" "	87.7	12.603	149.133	117041	80804.4	69.04	0.9683
" "	78.0	12.480	158.877	133471	99109.8	80.27	0.8660
" "	72.7	12.432	161.225	124814	103522.0	82.94	0.8083
" "	69.1	12.172	162.568	125223	104652.9	83.57	0.7701
" "	68.0	12.408	163.572	126387	105401.9	83.40	0.7588
" "	66.5	12.361	163.097	125986	104857.4	83.26	0.7412

TABLE XXIV.

Coefficients of Useful Effect for several heights of Speed Gate, and Velocities of Wheel, deduced from the Experiments made on the Seventy-two inch Swain Turbine Water-Wheel, in Boott Cotton Mill No. 5, in August, 1871.

Coefficients of Useful Effect : the Ratio of the Velocity of the Exterior Circumference of the Wheel to the Velocity due the head acting on the Wheel, being as given in the headings of the several columns.

Height of Opening Speed Gate, Inches.	0.60	0.61	0.62	0.63	0.64	0.65	0.66	0.67	0.68	0.69	0.70	0.71	0.72	0.73	0.74	0.75	0.76	0.77	0.78	0.79
13.00	0.765	0.771	0.776	0.782	0.788	0.793	0.798	0.802	0.808	0.813	0.818	0.822	0.827	0.830	0.832	0.834	0.835	0.835	0.834	0.833
12.00	0.773	0.780	0.786	0.791	0.796	0.802	0.807	0.811	0.816	0.820	0.824	0.828	0.832	0.835	0.837	0.838	0.839	0.838	0.837	0.835
11.00	0.775	0.781	0.786	0.792	0.795	0.802	0.807	0.812	0.817	0.821	0.825	0.829	0.831	0.833	0.832	0.832	0.830	0.827	0.824	0.819
10.00	0.782	0.788	0.794	0.800	0.805	0.810	0.815	0.820	0.824	0.824	0.828	0.831	0.831	0.831	0.830	0.829	0.828	0.824	0.820	0.811
9.00	0.779	0.785	0.790	0.796	0.802	0.808	0.812	0.816	0.820	0.823	0.825	0.827	0.826	0.825	0.825	0.822	0.820	0.815	0.811	0.806
8.00	0.771	0.777	0.782	0.788	0.796	0.800	0.804	0.808	0.801	0.802	0.804	0.807	0.806	0.805	0.800	0.804	0.798	0.795	0.791	0.788
7.00	0.764	0.769	0.773	0.778	0.781	0.784	0.786	0.788	0.790	0.790	0.790	0.790	0.788	0.788	0.785	0.780	0.777	0.772	0.768	0.763
6.00	0.742	0.746	0.751	0.754	0.756	0.759	0.760	0.761	0.761	0.761	0.759	0.758	0.756	0.755	0.751	0.746	0.743	0.739	0.734	0.728
5.00	0.708	0.711	0.714	0.717	0.719	0.721	0.722	0.722	0.723	0.723	0.719	0.718	0.715	0.715	0.713	0.710	0.707	0.704	0.701	0.695
4.00	0.654	0.658	0.661	0.664	0.664	0.666	0.669	0.669	0.668	0.667	0.665	0.663	0.660	0.657	0.655	0.652	0.648	0.645	0.641	0.637
3.00	0.556	0.559	0.581	0.583	0.585	0.586	0.585	0.586	0.586	0.583	0.582	0.581	0.578	0.573	0.571	0.568	0.563	0.559	0.555	0.550
2.00	0.810	0.414	0.423	0.422	0.471	0.469	0.468	0.465	0.460	0.456	0.451	0.446	0.441	0.434	0.428	0.425	0.412	0.404	0.395	0.385

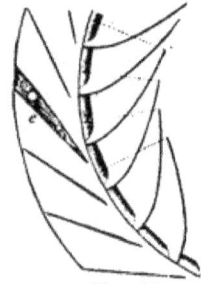

FIG. 44a.

The Swain Turbine

FIG. 44

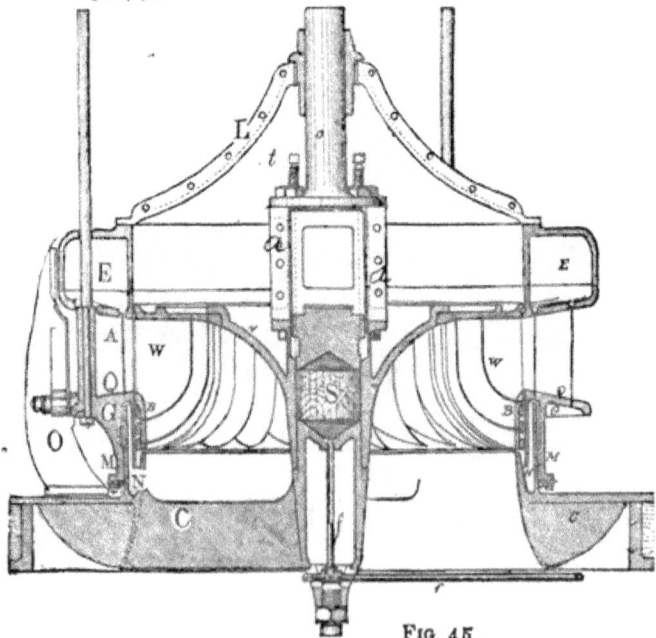

FIG 45.

69. *The Hercules.*—The Hercules turbine is also a mixed flow wheel, inward and downward. The entrance surface A A, Fig. 47, is conical, and divided horizontally by several partitions, so as to be equivalent to several wheels superimposed, but cast and working as one. The partitions are for the purpose of securing a good efficiency with partial gate opening; indeed, the best efficiency was for partial opening, from which it is inferred that at full opening the streams within the wheel interfered with each other, and so prevented their producing the best effect. The passages are so curved that the inner filaments passed downward and issued nearly radially, while the outer ones escaped at B more nearly axially. A cylindrical gate surrounded the wheel and was opened by being raised, similar to the Fourneyron style, or as was done in the Francis wheels, Figs. 19 and 31. The guides were outside the gate and stationary, as in Fig. 31, and divided by horizontal partitions. In 1883 a "Hercules" was tested at the Holyoke testing station. The wheel was of the following dimensions:

DIMENSIONS OF THE " HERCULES" TURBINE.

Mean external radius, ... $2r_m = 36$ in.
Terminal angle of centre line guides, $\alpha = 14\frac{3}{4}°$
Initial angle of buckets. . . . . $\gamma_1 = 98°$
Terminal angle of buckets (say) . . $\gamma_2 = 14°$
Measured least section of guides, . $K' = 4.752$ sq. ft.
" " terminal section of
buckets. . . . . . $k_2' = 7.925$ sq. ft.
Number of buckets, ... $N = 24$
" " guides, ... $N_1 = 17$

The first test showed such remarkable results in regard to efficiency, that it was followed by two other tests, all of which seemed to confirm each other. The following are from the third test :

TABLE XXV.

Abstract of (Third) Test of a 60-inch Hercules Turbine.

GATE OPENING.	No. of Revolutions per minute, n.	Total Head, Feet. H.	Volume of Water per sec., cu. ft. Q.	Brake Power. $H.P.$	Brake Efficiency, per cent. E.	Ratio. $\frac{u' \, r}{\sqrt{2\,g\,H}}$
1.000	Still	16.82	0	0	0	0
	129.10	16.94	89.66	145.59	84.55	0.614
	135.33	16.95	89.00	146.02	85.38	0.644
	140.62	16.96	88.33	145.72	85.80	0.669
	144.80	16.98	87.79	143.88	85.14	0.688
0.806	123.00	16.88	79.38	129.71	85.39	0.586
	130.25	16.88	78.63	131.01	87.07	0.621
	137.00	16.92	78.00	130.28	87.07	0.652
	143.00	16.94	77.38	129.01	86.82	0.680
0.647	126.50	17.09	67.25	111.04	85.22	0.599
	134.67	17.15	66.52	111.65	86.33	0.637
	140.40	17.16	65.72	109.55	85.69	0.664
0.489	115.60	17.25	55.54	85.27	78.50	0.545
	123.00	17.22	54.80	85.48	79.90	0.581
	130.00	17.28	54.07	84.79	80.05	0.613
0.379	112.75	17.61	45.20	65.29	72.36	0.526
	122.67	17.65	44.59	65.06	73.06	0.572
	129.40	17.66	43.76	63.89	72.93	0.603

The hydraulic efficiency was probably about 87 or 88 per cent.

Analysis of the "Hercules" Wheel.—Making $r_1 = r_2 = r_m = 18''$, $\alpha = 14\frac{3}{4}°$, $\gamma_1 = 98°$, $\gamma_2 = 14°$, $\mu_1 = 0.10$, $\mu_2 = 0.05$, gives

$$E = 0.8998 = 90 \text{ per cent. nearly.}$$

With the same data except $\mu_1 = \mu_2 = 0.10$ gives

$$E = 0.8803 = 88 \text{ per cent.,}$$

which exceeds the brake efficiency only one per cent., and indicates that the resistances are low and about 0.10 for μ_1 and μ_2.

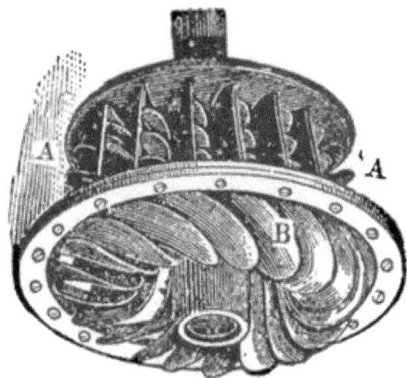

FIG. 46

70. *The Victor.*—The "Victor" turbine followed the Hercules in its introduction to the public. It is a mixed-flow turbine, the water entering radially inward at the circumference, thence discharging downward and outward. The whole body of the wheel, except that for the shaft and step, is occupied by the buckets, which are deep axially, thus giving great capacity for its size.

The water is regulated by two styles of gates: one is called "the register gate;" the other, "the cylindrical gate." The former opens the passages for the water by turning about the axis of the wheel, thus opening the passages their whole length and making the opening wider and wider as it is turned more and more, and at the same time gives direction to the water; the latter is a cylinder moving axially, and opens the passages their full width as the gate is raised. The latter is preferable when the water supply is variable and not sufficient to fill the passages at full gate, or when the work is variable; for a better efficiency is obtained with it at "partial gate." A view of the latter is shown in Fig. 48, for which we are indebted to the courtesy of the manufacturers, Stillwell-Bierce and Smith-Vaile Co., Dayton, Ohio.

TABLE XXVI.

RESULTS OF TESTS OF THE "VICTOR" TURBINE AT THE HOLYOKE TESTING STATION, WITH THE "CYLINDER GATE."

SIZE OF WHEEL AND GATE OPENING.	Head in Feet. H.	Revolutions of Wheel per minute. n.	Cubic Feet Water per minute. 60 Q.	Horse-power Developed by Wheel. $H. P.$	Percentage Useful Effect. E.
30-inch Full Gate.　..	17.51	168	4440	119.56	81.35
7/8　"　.....	17.82	163	3892	104.93	80.03
3/4　"　.....	17.95	163	3392	88.24	76.66
5/8　"　.....	18.10	155	2893	70.97	71.28
1/2　"　.....	18.20	159	2265	51.42	63.46
36-inch Full Gate.....	16.78	135	6106	158.18	81.80
7/8　"　.....	17.14	135	5422	141.58	80.71
3/4　"　.....	17.35	140	4708	118.22	76.68
5/8　"　.....	17.05	129	3982	91.62	71.50
1/2　"　.....	17.48	134	3202	66.87	63 30
39-inch Full Gate.....	14.66	116	6873	152.66	80.37
7/8　"　....	14.53	118	5920	129.41	79.80
3/4　"　.....	16.84	125	5517	135.56	77.40
5/8　"　.....	17.06	123	4695	108.22	71.67
1/2　"　.....	17.39	124	3856	81.00	64.07
48-inch Full Gate.....	13.23	91	10072	201.71	80.11
7/8　"　.....	14.36	89	9042	192.41	78.42
3/4　"　....	14.75	89	7869	165.23	75.34
5/8　"　.....	14.87	85	6744	132.76	70.06
1/2　"　.....	15.28	87	5526	100.66	63.09

NOTE.—In the above calculations "fractional gate" means "fractional water," regardless of the position of the gate of the turbine; that is, "three-quarters gate" means that the gate is closed to a point where the discharge of water is only three-fourths of what it is at full gate *under the same head.* This is nearly but not perfectly exact.

THE VICTOR TURBINE.

"CYLINDER GATE"

FIG. 46.

TABLE XXVII.

RESULTS OF TESTS OF THE "VICTOR" TURBINE AT THE HOLYOKE TESTING STATION, WITH THE "REGISTER GATE."

SIZE OF WHEEL. FULL GATE.	Head In Feet. *H.*	Revolutions per minute. *n.*	Horse-power. *H. P.*	Cubic Feet of Water per minute. 60 *Q.*	Percentage Useful Effect. *E.*
15-inch	18.06	368	30.17	990.19	.8932
	18.08	355	30.12	996.83	.8849
17½-inch	18.02	280	35.51	1164.60	.8960
	17.96	292	36.35	1197	.8950
20-inch	18.22	286	48.75	1660.17	.8532
	18.23	275	48.75	1660.17	.8528
25-inch	17.79	205.5	67.72	2362.72	.8530
	17.96	209	68.62	2356.54	.8584
30-inch	11.65	144.5	52.54	2751.87	.8676
	11.66	147.5	51.96	2755.09	.8564
35-inch	17.31	151.7	135.68	4895	.8489
	17.29	160	133.19	4806	.8497
40-inch	16.49	130	148.93	5789	.8253
	16.47	124	148.88	5816	.8221
44-inch	15.50	109.25	155.78	6643	.8003
48-inch	15.51	102	179.29	7456	.8202

It will be seen that the highest efficiencies, over 89 per cent., border very closely upon the figure which many writers reject without discussion. The engineer who made the tests has remarked, in effect, that the smallest wheels were too small for determining the efficiency with great accuracy. Even if these be excluded, the efficiencies are high.

71. *Pelton Wheel.*—The Pelton wheel, or, as it was formerly called, the " hurdy-gurdy" wheel, has been popular, especially in California, for utilizing the power of high waterfalls. It con-

sists of a series of curved double buckets, as shown in Figs. 48 and 49, attached to the circumference of the wheel. A is the nozzle, B the valve case, in Fig. 49a. The water impinges against the buckets at the partition and flows outward in opposite directions, discharging at the outside planes of the wheel. It is then an impulse wheel, or wheel of free deviation, with axial flow and segmental feed, and may be analyzed according to Article 30. For the limiting values of the angles, we have

$$\alpha = 0, \gamma_1 = 180, \gamma_2 = 0; r_2 = r_1; \mu_1 = \mu_2 = 0;$$

for which we find

$$V = \sqrt{2\,g\,H},$$

$$\omega\, r_1 = \text{velocity of circumference} = \tfrac{1}{2}\sqrt{2\,g\,H},$$

equation (36), which is half the velocity due to the head. The velocity of discharge will be, equation (34), $V_2 = 0$, and the efficiency will be unity. But these are impossible conditions in practice. The angle EBC must be less than 90°, so that the water will escape before it is struck by the succeeding bucket. Draw BD parallel to the line of the jet (or in the plane of the wheel), to represent the velocity of the wheel, and BC tangent to the bucket at its terminus, to represent the velocity of the water as it leaves the bucket relative to the bucket, and complete the parallelogram; then will EB represent the velocity of exit from the bucket relative to the earth. Considering the angle at A as zero, V_1 the velocity of the rim of the wheel $= BD = CE$, and V the velocity of the jet, then will the relative velocity of the water along the concave surface of the bucket be

$$V - V_1 = BC = ED = \mu\sqrt{2\,g\,H} - V_1.$$

If the velocity of the wheel be such that EB is perpendicular to the jet (or to the plane of the wheel), then

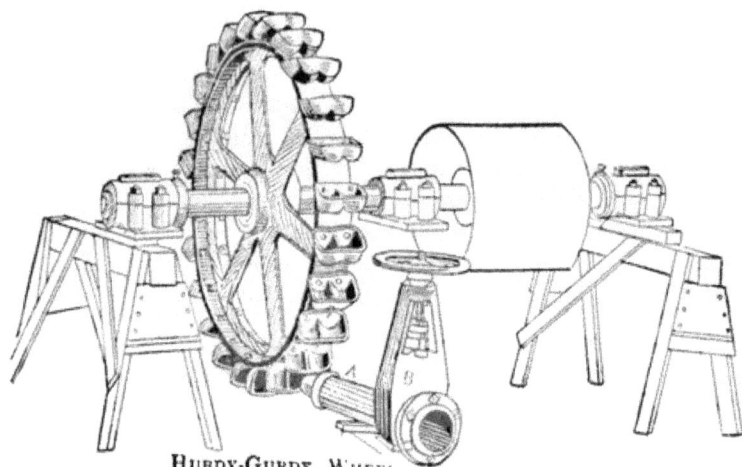

HURDY-GURDY WHEEL.

FIG. 45.

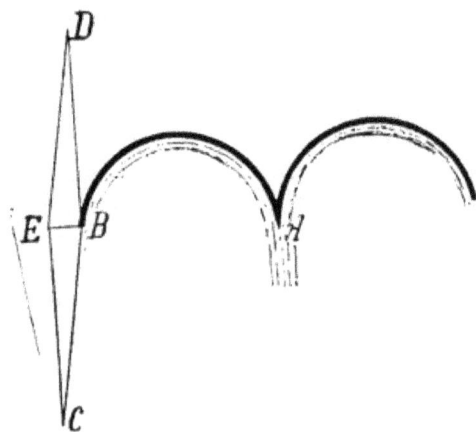

FIG. 49.

$$V_1 = BC \cos ECB = (V - V_1) \cos \gamma_2.$$

$$\therefore V_1 = \frac{\cos \gamma_2}{1 + \cos \gamma_2} V.$$

$$V_2 = EB = BC \sin \gamma_2 = \frac{\sin \gamma_2}{1 + \cos \gamma_2} V.$$

$$U = \tfrac{1}{2} \frac{\delta Q}{g} (V^2 - V_2^2).$$

$$E = \frac{U}{\delta Q H} = \mu^2 \frac{V^2 - V_2^2}{V^2}.$$

Thus, if $\mu = 0.98$, $ECB = 20°$, then
$V = 0.98 \sqrt{2 g H}$, $V_1 = 0.476 V$, $V_2 = 0.18 V$, $E = 0.928$,
or nearly 93 per cent.

If the velocity of the wheel be half that of the jet, then

$$EC = V_1 = \tfrac{1}{2} V;$$
$$BC = V - V_1 = \tfrac{1}{2} V;$$
$$\therefore BC = EC;$$
$$\therefore V_2 = EB = 2 \sin \tfrac{1}{2} C;$$

and U and E the same as that given above. When the angle C is small—say less than $25°$, $2 \sin \tfrac{1}{2} C = \sin C$ with sufficient accuracy for this case, and the results will be very nearly the same as those given above, so that for the conditions assumed above, the theoretical efficiency will be about 93 per cent. for this case. The efficiency will be nearly the same for a speed several per cent. above or below that for the maximum.

If the angle $C = 30°$, and the half angle at A also $30°$, and the speed be one-half the component of the velocity of the jet, and $\mu = 0.96$; then

$$V = 0.96 \sqrt{2 g H},$$
$$V_1 = \tfrac{1}{2} V \cos 30° = 0.433 V,$$
$$BC = V - V_1 = 0.433 V,$$
$$V_2 = 2 BC \sin 15° = 0.224 V;$$
$$\therefore E = 0.92 (1 - 0.224^2) = 0.874.$$

In these computations no allowance has been made for the friction in the feed pipe, nor for imperfect working at the junction of the buckets, nor for loss due (if any) to imperfect discharge. These all operate to reduce the efficiency.

In a paper read by Mr. Hamilton Smith, Jr., before the American Society of Civil Engineers, February 6, 1884, a test is reported in which the efficiency was given as 87.3 per cent. with a speed of 0.51 $\sqrt{2\,g\,H}$, under a head of 386 feet, wheel 6 feet in diameter. He states that the wheel carries over a large amount of water. This efficiency is remarkable, but if it be admitted that there were incidental errors in the test, it still shows that the efficiency was high, and Mr. Hamilton infers that it was certainly as high as 85 per cent.

A small "hurdy-gurdy" was tested at Stevens' Institute by some students in 1891, for which the efficiency was found to be less than 70 per cent.

The wheel is especially commendable for great heads: it is simple in construction, durable, efficient, easily managed, and easily repaired. It was invented by a village carpenter, who, after reading Francis' *Lowell Hydraulic Experiments*, and residing near a fall of water, made a Prony friction brake and a weir, and by continuous experiments arrived at the particular form and setting of buckets which he adopted as the best. The juncture *A* in the actual bucket is back of the line, joining the extremities through *B*.

72. *Poncelet Wheel.*—The Poncelet wheel is the invention of him whose name it bears. In accordance with hydraulic principles, the inventor changed the plane float wheel, which had an efficiency of about 16 per cent., to one with curved buckets, Fig. 50, raising the efficiency to 60 or 70 per cent. The water enters the buckets tangentially, moving up the concave side with a diminishing velocity since it works against gravity, until it ceases to ascend, when it descends and leaves the bucket with a backward

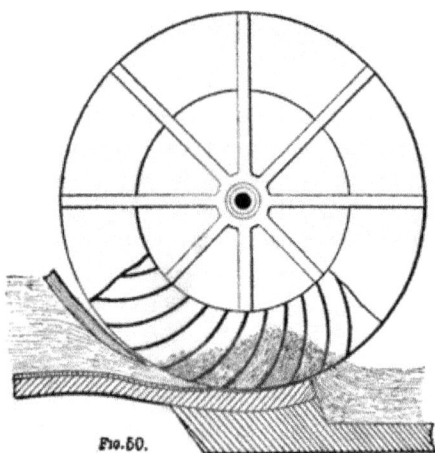

Fig. 50.

velocity in reference to the bucket, its velocity in reference to the earth depending upon the velocity of the wheel. The energy lost in its work against gravity is restored by gravity during descent, neglecting friction in the buckets, so that it is only necessary to consider the energy of the jet at entrance and quitting. It is virtually a turbine of "free deviation," with segmental feed. Considering the limiting case, in which the terminus of the bucket is tangent to the outer circumference, and that the water enters tangentially, then

$$\alpha = 0, \gamma_1 = 180°, \gamma_2 = 0, \mu_1 = 0 = \mu_2, r_1 = r_2,$$

then

Equation (93), $V = \sqrt{2 g H},$
"   (94), $v_1 = V - \omega r,$
"   (97), $\omega r = \tfrac{1}{2} \sqrt{2 g H} = \tfrac{1}{2} V,$
       $\therefore v_1 = \tfrac{1}{2} V,$
"   (34), $V_2 = 0,$
"   (101), $U = \tfrac{1}{2} M V^2,$
"   (102), $E = 1.$

But these conditions cannot be realized in practice. The terminal angle of the bucket has an angle of 15° or more with the circumference, or $\gamma_1 = 165°$. The angle of the guide, or chute, will be 20° or 25° or even more, and $\gamma_2 = 15°$ or more, and $\mu_1 = 0.10$ or more. These would give an efficiency of less than 90 per cent. There may be a further loss by water escaping between the wheel and apron below the wheel, and there may be an imperfect action of a finite stream; since the bucket may strike the stream, and the stream strike the crown, thus reducing the theoretical efficiency. Actual tests give

$$E = 0.55 \text{ to } 0.70,$$

when

$$v = 0.55 V, \text{ about.}$$

73. In the preceding tests we have given the highest efficiencies, but it should not be inferred that these high figures are attained in the majority of cases. In Europe, Rittenger tested eight turbines, parallel flow, mean radii from 6 to 11 inches, having various vane angles, under heads varying from about 6 to 18 feet, for which he found efficiencies varying from 0.63 to 0.71. It may be observed that tests of American wheels in America— or we might say in Massachusetts—have given higher efficiencies than tests in Europe, and the question has been raised, if the difference may not be due to the different methods of measuring the quantity of water; but the difference is not accounted for in this way. There appears to be essential differences in the wheels in favor of the American types.

In designing, it is not advisable to use the highest attainable figures for the efficiency, for when tested the wheels are supposed to be in their best condition, well lubricated and bearing perfect, and after long service, some parts may be so worn as to make the wheel less efficient; hence 70 per cent., or certainly not to exceed 75 per cent., should be assumed. In ordinary practice, with variations of speed above or below that which would produce a maximum, with wheels not constructed with great care, and with a lack of proper attendance they may fall below 60 per cent.

CLASS-ROOM EXERCISE.

(The following exercise was conducted by the author with a class.)

Design a turbine to utilize the power of a stream having an available fall of 16 feet.

74. *To find the power of the fall.*

Let

δ be the weight of a cubic foot of water,

Q, the volume of water falling per second,

H, the height of the fall;

then the energy of the fall per second will be

$$\delta \, Q \, H,$$

and the horse-power,

$$H.P = \frac{\delta \, Q \, H}{550}.$$

The weight of a cubic foot of water depends upon its temperature, latitude and elevation of the place; but in the use of water wheels the temperature will vary so little from 60° or 70°, and the latitude and elevation will have so small an effect, we will use

$$\delta = 62.4.^*$$

---

* If T be temperature, degrees Fah.,

" l, the latitude of the place,

" e, the elevation above sea level,

" $\delta_0 = 62.375$, the weight of a cubic foot of water at its maximum density, 39.1°, at the level of the sea at latitude 45°, then

$$\delta = \delta_0 \, (1 - 0.0002) \, (T - 39.1°) \, (1 - 0.000256 \cos 2 \, l) \left(1 - \frac{2 \, e}{r}\right)$$

If r be the radius of the earth at the place,

r_0 "   "   "   " 45°, then

$$r = r_0 \, (1 + 0.00164 \cos 2 \, l),$$
$$r_0 = 20,892,200 \text{ feet.}$$

In case of accurate tests, the proper allowance for these changes may be made.

The head over the crest of the weir will vary from zero at the crest to some finite value depending upon the depth of water over the crest. To deduce a formula for this case, first assume that the water flows through an orifice $B\,D$, Fig. 51, in which $C\,D$ is the breadth, $B\,C$ the depth, and A the free surface. Let E be a rectangular element $=d\,x\,d\,y$, in which y is vertical and x horizontal, and y the depth of E below the free surface; then

$$d\,Q = \mu\ \sqrt{2\,g\,y}\ .\,d\,y\,d\,x.$$

Let

μ be a coefficient of discharge,

h_2, the depth $C\,A$,

h_1, "   " $B\,A$,

l, the breadth $C\,D$:

then

$$Q = \mu\ \sqrt{2\,g}\ \int_0^b \int_{h_1}^{h_2} \sqrt{y}\,d\,y\,d\,x = \frac{2}{3}\mu\ \sqrt{2\,g}\,l\ \left(h_2^{\frac{3}{2}} - h_1^{\frac{3}{2}}\right).\ (a)$$

If the upper surface of the orifice be free—in other words, if the orifice be a weir, as in Fig. 52—it has been found by experiment that h_1 may be taken as zero when $h_2 = B\,D$, measured from the level $A\,B$ of the water several feet above the weir down to the crest of the weir D. The coefficient μ varies between 0.60 and 0.64, depending upon the depth $C\,D$ over the weir and $B\,F$ the breadth; and if one were gauging a stream where great accuracy was essential, it would be necessary to resort to tables to determine the exact coefficient to be used (D'Aubisson, *Hydraulics*, or Weisbach, *Hydraulics*), or resort to Francis' formula as given in *Lowell Hydraulic Experiments*, which is

$$Q = 3.33\ (l - 0.1\ n\ h)\ h^{\frac{3}{2}}.$$

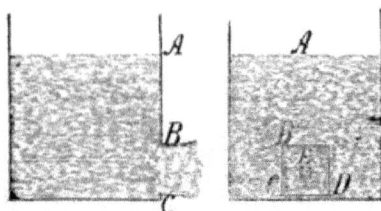

FIG. 51.

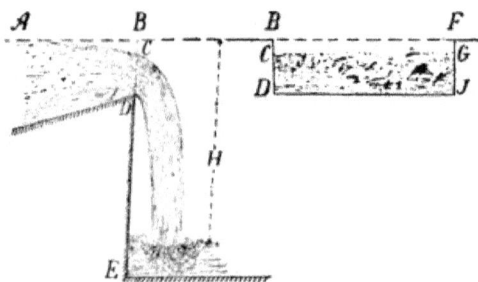

FIG. 52.

in which n is the number of contractions (*two*, if both ends are contracted; *one*, if only one end; and zero is neither end). If

$$n = 0,$$
$$Q = 3.33 \, l \, h^{\frac{3}{2}},$$

to which the preceding formula will reduce if $h_1 = 0$ and $\mu = 0.622$. Streams vary continually in the quantity of water discharged, so we will use the smaller coefficient, or $\mu = 0.60$, and our equation becomes

$$Q = 3.21 \, l \, h^{\frac{3}{2}}. \tag{b}$$

Let a weir be constructed of boards having bevelled edges—the sharp edges being placed up stream—and by measurements suppose it be found that

$$l = 6 \text{ feet},$$
$$h_2 = 1.83 \text{ feet}.$$

The height of the surface may be accurately found by a " hook-gauge," a device invented by Mr. Francis, in which a hook is submerged and then gradually raised until the point is just visible in the surface. In this way the height of the surface may be found with much accuracy to a small fraction of an inch. We have

$$Q = 3.21 \times 6 \times (1.83)^{\frac{3}{2}} = 48 \text{ cubic feet}$$

per second, nearly, and sufficiently accurate for our present purpose. The theoretical power of the fall will be

$$H. P. = \frac{48 \times 62.4 \times 16}{550} = 87.13$$

horse-power. If the turbine has 75 per cent. hydraulic efficiency, it will have 65.35 horse-power; and if the frictional resistances be 3 per cent. of this power, it will have at the brake 63 horse-power. From this we can judge whether the stream will supply the required power.

75. *To determine the diameter of the wheel.*—We choose for this exercise a radially outflow wheel. The greater the diameter of the wheel, the less will be its depth in order to discharge the given amount of water. There is no recognized rule for determining the diameter. The depth between the crowns of the Tremont wheel was about $\frac{1}{5}$ the outside diameter. If proportioned for space or to economize material, the depth would be more nearly equal to the radius. We will try a depth equal to $\frac{1}{4}$ the radius, r_2. Table VI., column 10, second case, gives

$$V_2^2 = 0.04 \times 2\,g\,H.$$
$$\therefore V_2 = 0.2\,\sqrt{2\,g\,H}.$$

The Tremont wheel gave

$$V_2 = 0.24\,\sqrt{2\,g\,H} = \frac{7.00}{28.8}\,\sqrt{2\,g\,H}, \text{ nearly.}$$

Assuming that the free opening at the outer rim (the circumference less the space occupied by the walls of the buckets) is $\frac{2}{8}$ of the outer circumference, and that the radial velocity at quitting is $0.2\,\sqrt{2\,g\,H}$, and depth $\frac{1}{4}\,r_2$, we have

$$\frac{2}{8} \cdot 2\,\pi\,r_2 \cdot \frac{1}{4}\,r_2 \cdot 0.2\,\sqrt{2\,g\,H} = 48;$$
$$\therefore r_2 = 2.6 \text{ feet, nearly}$$
$$= 31.2 \text{ inches.}$$

The object of this investigation is not to fix exactly the proportions of the wheel, but to determine approximately such proportions as may be previously desired. We may now assume, arbitrarily, the outer radius, and if not satisfied with the final result, recompute with another assumed radius.

We observe that the smaller the diameter, the greater will be the number of revolutions per minute, and the wheel may be proportioned to give the required revolutions; but the analysis is complex, as shown by equations (15a) and (16), page 8. By the aid of Table VI., a first approximation may be made; thus, if n be the number of revolutions per minute, then

$$\omega' = \frac{n \cdot 2\,\pi}{60} = \frac{n\,\pi}{30},$$

and from Table VI.

$$r_2 = 0.875 \ \sqrt{2\,g\,H} \cdot \frac{30}{n\,\pi}.$$

Thus, if $H = 16$ feet, $n = 100$ per minute; then

$$r_2 = 2.67 \text{ feet} = 32.04 \text{ inches};$$

but the coefficient 0.875 depends upon values we have not yet fixed. In several cases given in the preceding pages, the velocity of the initial rim to the velocity due to the head is between 0.62 and 0.70. We now assume a value near those found above, but otherwise arbitrarily,

$$r_2 = 30 \text{ inches.}$$

The width of the crown will now depend upon the inner radius, and no rule exists for determining this. In the Tremont wheel $r_2/r_1 = 1.23$. Rankine gives $r_2 = r_1 \sqrt{2} = 1.41\,r_1$. These give, for our problem, $r_1 = 24.4$ inches and 21.27 inches respectively; and for width of crown 5.6 and 8.73 inches respectively. The analysis on page 26 shows that, theoretically, the crown should be comparatively narrow for the outflow wheel, but from physical considerations the crown should be sufficiently wide to secure the full effect of the stream; but if too wide, friction and difficulty in proper feeding might be prejudicial. For the purpose of study, a wide crown will show more clearly all the peculiarities of the buckets than a very narrow one, especially in the graphical construction. Then assume $r_1 = 20$ inches, giving 10 inches for the width of crown. (The width of crown in the Tremont was nearly 9¼ inches.)

76. *Initial angle of the bucket.*—For reasons given previously, we make

$$\gamma_1 = 90°.$$

77. *Terminal angle of bucket.*—The discussion in Article 5, page 9, shows that γ_2 should be small for high efficiency. The smaller it is the greater must be the outer depth, and if the wheel is to be cast, the angle should not be too small. In the Tremont wheel, in which the walls of the buckets were of Russia sheet iron and fitted into place by special tools, this angle was 10 degrees. Other wheels mentioned in this work are made with larger angles. We will assume

$$\gamma_2 = 14°.$$

78. *Terminal angle of the guide vanes,* directrices, or distributors.—The proper value of α is discussed on page 16, and that the internal pressure at the gate should exceed that of an atmosphere it should be less than $45°$ for $\gamma_1 = 90°$. The smaller it is the higher the efficiency, although the gain in efficiency is small per degree of decrease of the angle. It may be seen in Table X., that for an increase of α from $21°$ to $23°$, the loss of efficiency was 0.23 of one per cent., or not far from $\frac{1}{16}$ of one per cent. per degree for those values. We will assume the fairly practical value,

$$\alpha = 30°.$$

79. *Buckets.*—There is no recognized rule for determining the number of buckets or their form. Francis' rule for the number is

$$N = 3 (D + 2),$$

where D is the outside diameter in feet; but he did not follow it in the construction of his wheel. He made $N = 44$, while the rule would give 30. In our example it gives $13\frac{1}{2}$, and 12 or 14 would be used; or if in the proportion of the diameters to the Tremont, about 26. The wheels tested at the Centennial Exposition had less buckets, page 41, than given by this formula. Circumstances pertaining to the ease and certainty of construction, or of obstacles entering the wheel, may have controlling influences. A case is related of a wheel choked and stopped by eels, but it is rare for a wheel to act as an eel-trap; but if there is danger of obstructions, it would be wise to make the buckets larger. We will try $N = 18$, and 16 guides. The thickness of the partitions will depend upon the mode of manufacture. If of sheet steel, they may be $\frac{1}{8}$ inch thick; if of bronze, less than $\frac{1}{4}$ of an inch; but if of cast iron, they should be thicker to secure sound castings, and some allowance may be made for imperfect setting of the cores. So many cores will be necessary that dry sand moulds should be used, in which case sound castings may be secured if $\frac{3}{8}$ inch thick and possibly if less; but to provide for contingencies, we will design them $\frac{7}{16}$ inch thick and bevel the initial ends to a sharp edge. With this data we make the first calculation. The data are

$r_1 = 20''$, $\gamma_1 = 90°$, $\mu_1 = 0.10$, $H = 16$ ft.
$r_2 = 30''$, $\gamma_2 = 14°$, $\mu_2 = 0.20$, $Q = 48$ cu. ft. per sec.
$t = \frac{7}{16}''$, $\alpha = 30°$. $N = 18$ buckets.

RESULTS.

$M^2 =$	0.608	N^2	$= 0.052$
Ang. vel. $\omega' =$	10.92	E	$= 0.737$
Rev. $=$	1.738 per sec.	K	$= 0.127$ sq. ft.
$=$	104.28 per min.	k_1	$= 0.254$ " "
$V =$	21.016 ft.	k_2	$= 0.089$ " "
$v_1 =$	10.501 "	y_1	$= 0.466$ ft.
$v_2 =$	29.824 "	y_2	$= 0.511$ "
$a_1 =$	0.545 "	θ	$= 77° 10'$
$a_2 =$	0.722 "	$H P$	$= 64.31$, hydraulic
$V_2 =$	7.403 "		horse-power of the wheel.

80. *Form of the bucket.*—If the bucket is to be described with several arcs of circles, proceed as in Fig. 19. Let F, Fig. 19

or 53 be the terminus of a bucket, draw the radius $F O$, and
lay off $F c$, making an angle of 14° with $F O$. Choose a convenient point c_1 on $F c$, within but not far from the inner arc of
the crown for a centre, and with c_1 F as a radius, describe an arc.
$F f_2$, over about one fourth the width of the crown; then with
a shorter radius, $f_2' c_2$, where c_2 may be still within the inner arc
of the crown, describe $f_2 f_3$; then with c_3 the arc $f_3 f_4$; then
with c_4 complete the arc. If the initial angle γ_1 be 90°, the
centre c_4 must be on the tangent to the inner circumference at
G, and if it does not come out that way the correction should be
made by trial. The arcs through f_2, f_3, and f_4, with O as a centre,
may first be described dividing the crowns into four equal widths,
or any other proportion if desired.

It was found in the study of the Tremont turbine that if the
vane angles were used in the formulas and the depth y_2 be computed from the equation

$$(2 \pi r_2 \sin \gamma_2 - N t) v_2 y_2 = Q,$$

that y_2 was found to be too large. To find a value more nearly
the practical one, draw the tangent $a b'$, Fig. 53, and through u
the middle of the arc $F d$, a perpendicular $p r$ to $F g$. it will be
found that p is a point of tangency of a line parallel to $F g$.
The terminal angle γ_2 at u will be the same as the vane angle at
a, or 14°; hence the computed mean velocity at that point will
be $v_2 = 29.82$. Assume 0.85 as the coefficient of contraction,
although in the Tremont wheel it was 0.88, and would probably
be larger in this wheel, since γ_2 is larger; but it is better to make
the outer depth a little too large than too small, for if too small
eddies might be formed at the initial end of the bucket. Measuring on the drawing, we find

$$r p = 0.2847 \text{ ft.};$$
$$\therefore 29.82 \times 18 \times 0.85 \times 0.284 y_2 = 48;$$
$$\therefore y_2 = 0.378 \text{ ft.},$$

which value we use for the outer depth between the crowns.

81. *To find the form of the crowns.*—Let the arc A u, Fig. 53, be divided into (say) five equal parts, measured by trial, $A B = B C = C D$, etc. Describe arcs through B and C, etc., with the centre of the wheel as centre, and let

Radius at $\quad A$ be r_1.
" $\quad$ " $\quad B$, ρ'.
" $\quad$ " $\quad C$, ρ'', etc.
Velocity at A, v_1.
" $\quad$ " $\quad B$, v', etc.
Angle between the arcs at A, γ_1,
" $\quad$ " $\quad$ " $\quad$ " B, γ', etc.
Width of bucket at $A = a_1 = a' a''$,
" $\quad$ " $\quad$ " $\quad$ " $B = a' = b' B b''$,
" $\quad$ " $\quad$ " $\quad$ " $C = a'' = c' b c''$, etc.
Depth between the crowns at A, y_1.
" $\quad$ " $\quad$ " $\quad$ " B, y', etc.

Then for N buckets

$$N . a_1 y_1 . v_1 \sin \gamma_1 = Q.$$
$$N a' y'. v' \sin \gamma' = Q.$$
$$\therefore y' = \frac{a_1 v_1 \sin \gamma_1}{a' v' \sin \gamma'} y_1.$$

(c)

But $\dfrac{a_1}{r_1} = \dfrac{a'}{\rho''}$ very nearly, and if the thickness of the bucket be neglected, it will be exact—consider it exact : then

$$y' = \frac{r_1 v_1 \sin \gamma_1}{\rho' v' \sin \gamma'} y_1,$$

(d)

or,

$$y' = \frac{r_2 v_2 \sin \gamma_2}{\rho' v' \sin \gamma'} y_2.$$

We will seek to produce a practically uniformly increasing velocity from A to u. Then the increase from A to B will be

$$(v_2 - v_1) \frac{A B}{A u},$$

and the velocity at B will be

$$v' = v_1 + (v_2 - v_1) \frac{A\,B}{A\,u}. \qquad (J)$$

And if $A\,u = \mathbf{5}\,A\,B$, then

$$v' = \tfrac{4}{5}\,v_1 + \tfrac{1}{5}\,v_2 = 14.36.$$

Similarly at C,

$$v'' = \tfrac{3}{5}\,v_1 + \tfrac{2}{5}\,v_2 = 18.23.$$

TABLE XXVIII.

By Measurement on the Drawing.		From Computation as above.	From Equation (141).
$\gamma_1 = 90°,\ r_1 = 20$ inches		$v_1 = 10.50$ feet	$y_1 = 0.466$ feet
$\gamma' = 63°,\ \rho' = 23$ "		$v' = 14.37$ "	$y' = 0.332$ "
$\gamma'' = 44°,\ \rho'' = 25.6$ "		$v'' = 18.23$ "	$y'' = 0.304$ "
$\gamma''' = 30°,\ \rho''' = 27.6$ "		$v''' = 22.10$ "	$y''' = 0.331$ "
$\gamma'''' = 20°,\ \rho'''' = 29.0$ "		$v'''' = 25.96$ "	$y'''' = 0.408$ "
$\gamma_2 = 14°,\ r_2 = 30.0$ "		$v_2 = 29.82$ "	$y_2 = 0.511$ "

The values of y_1, y', y'', etc., will be the depths of the buckets at A, B, C, etc. Since these will be the depths at all points in the circumference of the arcs passing through those points respectively, draw a radius $G H$ and prolong the arcs to an intersection with this radius. On a line $A\,u'$, Fig. 54, equal to the width of the crowns, lay off the divisions of the radius, and through those divisions erect ordinates $y_1 = 0.466$, $y' = 0.332$, $y'' = 0.304$, etc., and trace a smooth curve through their extremities—these will be the form of the crowns for an indefinitely narrow bucket.

But for a bucket of finite width as we have in practice, the normal widths should be measured. These widths will be strictly a curve passing through the points A, B, C, etc., cutting normally the traces of an indefinite number of buckets between $G F$ and $a'' d$, as shown by the dotted line through D, but it will be sufficiently exact to consider the lines as straight. With dividers find by trial the shortest distance $B\,b'$ and $B\,b''$, and similarly for all the points C, D, etc. We find

TABLE XXIX.

Distances.	Velocities as found above.	Hence.
$a'\ A\ a'' = 0.556$	$v_1 = 10.50$	$y_1 = 0.466$
$b'\ B\ b'' = 0.572$	$v' = 14.37$	$y' = 0.331$
$c'\ C\ c'' = 0.492$	$v'' = 18.23$	$y'' = 0.304$
$d'\ D\ d'' = 0.396$	$v''' = 22.10$	$y''' = 0.311$
$e'\ E\ e'' = 0.328$	$v'''' = 25.96$	$y'''' = 0.320$
$p\ r = 0.284$	$v_2 = 29.82$	$y_2 = 0.322$
		$\dfrac{y_2}{0.85} = 0.378$

The values of y', y'', etc., are found from the equations

$$a'\ A\ a'' \cdot y_1 \cdot v_1 = b'\ B\ b_2 \cdot y' \cdot v' = c'\ C\ c_2 \cdot y'' \cdot v'' = \text{etc.} = \frac{Q}{N} \cdot$$

The crowns are constructed with these values in the same manner as above, and are shown in Fig. 54a.

It will be seen that the depths from the initial end to more than half the length are nearly the same for a finite stream as for an infinitesimal one, but that beyond the middle the depths are less for the latter, and this corresponds more nearly to what they should be in practice.

If the crowns are plane and parallel $y_1 = y' = y''$, etc., and the normal widths will be inversely as the velocities along the bucket; hence

$$b'\ B\ b_2 = a'\ A\ a''\ \frac{v_1}{v'} = \frac{10.50}{14.37}\ a'\ A\ a'' = 0.730\ a'\ A\ a''\ ;$$

$c'C'c_2 = 0.575\ a'\ A\ a''\ ;\ d'D\ d_2 = 0.475\ a'\ A\ a''\ ;\ e'\ E\ e_2 = 0.404$, etc.

In laying these off in Fig. 55, the centre line $A\ B\ C$ has been retained, and also the normals $B\ b'$, $C\ c'$, etc., and with the excess $B\ b_2 = b'\ B\ b'' - B\ b'$, an arc is described with B as a centre, and similarly at C, D, E, etc., and a curve $a''\ b_2\ c_2\ d_2$ traced tangent to the arcs. This forms, substantially, a back vane, the form of which will depend upon the law governing the velocity of the water along the bucket.

Fig. 55a shows the form of back vane when the velocity in the bucket is uniform from a'' to x, and equal to 10.50 feet; and from x to exit increasing to 29.82 feet.

82. The pressure at the entrance into the wheel will be given by equation (143), making $\rho = r_1$, $k = k_1$, then

$$p_1 = p_a + \delta h_1 + \tfrac{1}{2}\left[-1 - \mu_1 + \tfrac{1}{4}\mu_2\right]\frac{\delta}{g} \cdot \frac{\omega^2 r_1^2}{\sin^2 120°} = 2894.$$

The pressure at exit will be

$$p_2 = p_a + \delta h_2.$$

If the wheel be submerged one foot, then

$$p_2 = 2116 + 62.4 \times 1 = 2178.4 \text{ pounds per square foot.}$$

83. *To find the path of the water* in reference to the earth. The path is determined on the supposition that the velocity of the water is uniformly increasing as it passes through the wheel, as given in Table XXIX., on page 127. but is uniform in passing from A to B, B to C, etc., and then proceeding as in Article 50. The result is shown by the line $J K$, Fig. 53.

84. The diameter of the shaft for ample security may be found from the equation

$$d = \sqrt[3]{100\,\frac{H P}{n}} = \sqrt[3]{\frac{64}{1.04}} = 4 \text{ inches}$$

nearly, and $3\tfrac{3}{4}$ inches will be safe.

The wheel must be so secured to the shaft that it will not get free nor even slip. It may be secured by a shrunken band, as in Fig. 19, or by a key, as in Fig. 31. A key seat requires the cutting away of some of the material of the shaft, and at this point it may be advisable to make it 4 inches, even if it be less for the remaining part.

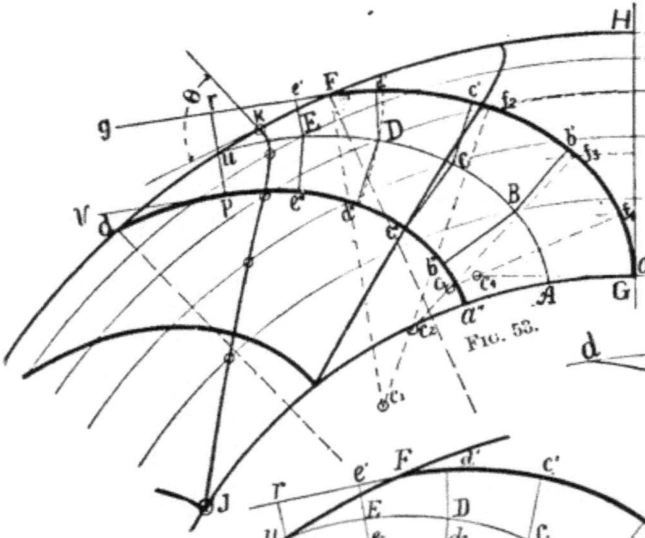

FIG. 53.

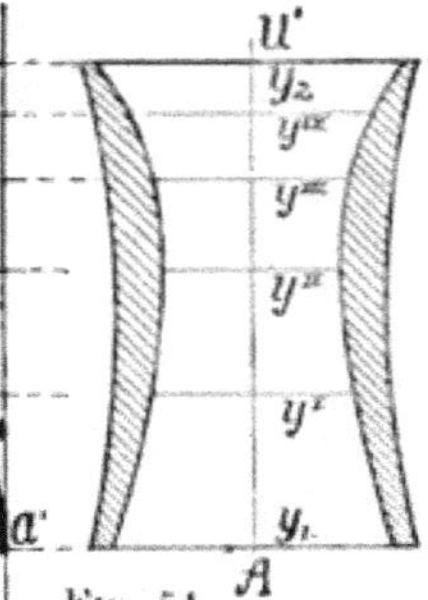

FIG. 54.

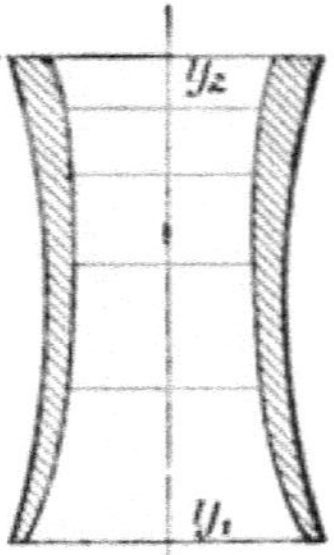

FIG. 54a.

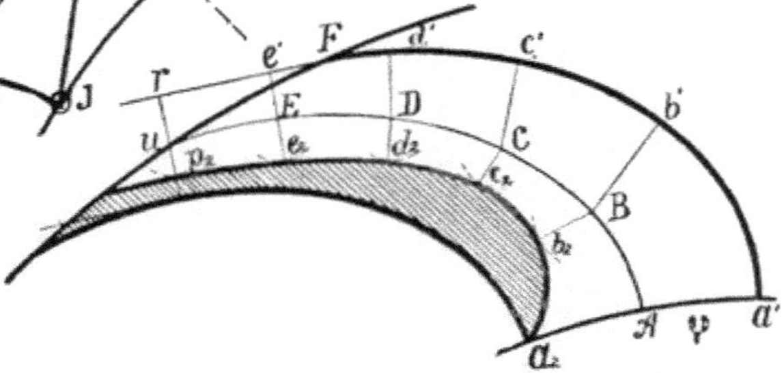

FIG. 55.

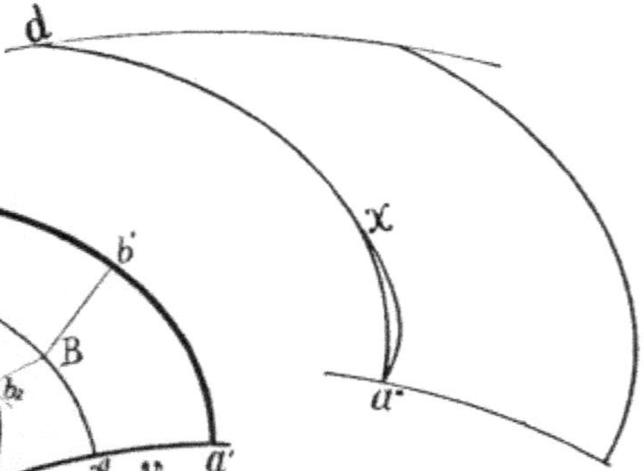

FIG. 55a.

85. Turbines are now frequently mounted on a horizontal shaft, producing practically the same efficiency as if vertical, and frequently are in pairs, the wheels facing each other, so as to relieve the shaft bearings of end or axial pressures.

The evolution of the modern American wheel seems to have begun with the "Swain" wheel, which was being developed from about 1862 to 1875. This in-and-axial flow wheel in its final form had a high efficiency, as shown in the test already given, and its construction was comparatively cheap. The wheel was cast solid, with less and deeper buckets than the Fourneyron and wider openings for discharge, and produced at a cost of, say, $\frac{1}{4}$ to $\frac{1}{2}$ of that of the built-up wheels of Boyden and Francis, and yielding a higher efficiency. Then followed the Risdon, with its reputed 87 per cent. efficiency, then the Hercules (1876) with a still smaller diameter and still deeper buckets and large discharge, cast solid, and yielding a "test" efficiency of 87 per cent.; and this was quickly followed by the "Victor" on the same general plan, but with different buckets and gate, and yielding as high if not higher efficiency. Other wheels of good repute have been produced during this evolution. The Leffel wheel, in which the inward and downward passages were separated by a diaphragm, and the "Humphrey" wheel, herein described, and still others, in all of which more than 80 per cent. efficiency has been delivered, so that the selection of a good wheel by purchasers must be made on other grounds. The capacity of a wheel in producing a given power is one of these elements.

The following is an approximate comparison of the wheels as they have usually, or, perhaps, it is safer to say as they were formerly, made, for the proportions of any one of them may be changed, and before this time may have been so changed as to produce different figures. The comparison is for a fall of 24 feet and diameter of wheel of about 36 inches.

TABLE XXX.

	Cu. ft. water per sec.	H.P.
Boyden-Fourneyron	30	65
Risdon.............................	50	115
Swain	100	220
Hercules............................	108	235
Leffel-Samson......................	110	240
Victor	111	241

According to these figures, the purchaser must choose his wheel upon other considerations than mechanical efficiency or economy of space.

86. *Niagara Wheel.*—The Niagara Falls Power Co. propose to utilize some 100,000 horse-powers of the Falls of Niagara. For this purpose a short canal or bayou 250 feet wide and 12 feet deep a mile or more above the crest of the falls, excavated for this purpose, conducts the water from Niagara River to vertical shafts, in the lower ends of which are placed, or are to be placed, turbines, which will have a head of about 136 feet. The tail races conduct the escaping water to a tunnel 7000 feet long, 21 feet high and about 18 feet wide, discharging into the river a short distance below the falls.

In this country most turbines are made from fixed patterns of definite sizes to suit average conditions, are turned out in quantities, and kept in stock, like store goods, to be purchased when wanted; but in Europe they are more generally made to order, designed for the particular place and conditions. So when the Niagara commission called for plans of wheels, the American manufacturers submitted their trade catalogues, and the European manufacturers submitted special designs. Among the designs considered was the American twin turbine on a horizontal shaft, with belt or rope transmission, but the difficulties encountered were so great that such plans were finally abandoned, and the design of Messrs. Faesch & Piccard, of Geneva, Switzerland, was accepted; the plan of which is shown in Fig. 56, for which the author is indebted to the courtesy of Coleman Sellers, E.D., Prof'r of Practical Eng'g Stevens Institute and Pres. and Chief Eng'r of The Niagara Falls Power Co. The water passes from the lower end of the penstock A, Fig. 57, into the casing B B, thence through the distributers b b b, three of which are at the upper end of the case and three below; thence through the buckets a a a. The wheel is divided into six elementary wheels by transverse discs, three of which are above where the water enters and three below. The gate c c is a cylinder exterior to the wheel, and opens the wheel passages by being moved downward. The shaft C passes through the central part of the case, to the former of which the crowns of the wheel are firmly secured. The upper crown is solid, but through the upper end of the case are openings d d, through which water may pass into the space between the upper end of the case and the upper crown of the wheel, which, acting by upward pressure, supports a part, or all, of the weight of the wheel, shaft, and attachments. The lower end of the case is solid, but the lower crown has openings h h, so that any water entering the space above it may readily escape so as not to produce a downward pressure upon the wheel. The lower end of the case is supported by three rods g g extending through the case, two of which are shown in the figure. The gate is regulated by a delicate and efficient regulator, so that the deviation from the velocity desired is less than 4 per cent. when the load is increased or decreased by 25 per cent.

The figures in Fig. 56 were used in the construction of the wheel. The wheel was made of bronze, and the buckets, partitions, and crowns immediately above and below the buckets are solid in one casting.

The terminal angle of the guides is given as 19° 6′, and of the buckets 13° 17½′, both of which, being comparatively small, are favorable to economy, and the comparatively small initial angle of the buckets, 69° 20′ (or less), makes it a wheel of considerable pressure.

The principal data are :

External diameter, ...	$2r_2 = 6$ ft. 3	in.
Internal "	$2r_1 = 5$ " 3	"
Width of crown, ...	6	"
External diameter of distributing chamber,	5 " 2⅞	"
Internal " " "	4 " 4	"
Clearance of wheel, ...	$\frac{1}{16}$	"
Width of distributing chamber, . . .	5$\frac{7}{16}$	"
Diameter of penstock, ...	7 " 6	"
Number of buckets, ...	$N = 32$.	
Number of guides, ...	$N' = 36$.	
Depth of each chamber a, ...	3⅘	"
Clear depth of six chambers $a\ a\ a$, . .	1.81 ft.	
Thickness of the horizontal partitions, each,	1	"
Terminal angle of guides, . .	$a = 19^{\circ}$ 6′	
Initial angle of buckets, about, . .	$\gamma_1 = 69^{\circ}$ 20′	
Terminal angle of buckets, . .	$\gamma_2 = 13^{\circ}$ 17½′	
Total head, about, ...	$H = 136$ ft.	

In regard to the head some writers have given it as 136 feet, while others have called it 140 feet. It is not necessarily constant, varying somewhat with the height of water in the river. Clemens Herschel, one of the engineers of the company, in *Cassier's Magazine*, of July, 1895, says : "The wheels will discharge 430 cubic feet per second, and acting under 136 feet head from the surface to the centre of the wheel, will make 250 revolutions per minute ; at 75 per cent. efficiency will give 5,000 horse-power." Mr. Stetson, an officer of the company, in the same magazine, speaks of 140 feet. If the wheel discharges 430 cubic feet, the velocity in the penstock will be between 9 and 10 feet per second, which will be equivalent to more than one foot head, so that the effective head will be over 137 instead of 136. The head at the lower end of the wheel will be about 10 feet greater than that at its upper end, and the mean head will be a little lower than to centre of the wheel. We will use for computation

$$H = 138 \text{ feet.}$$

The initial angle of the buckets is marked as $110° \, 40'$, the supplement of which in our notation is $\gamma_1 = 69° \, 20'$. But this is the mean angle between the face and back of the vane. We made a computation for the efficiency and speed using this angle; but it made the efficiency very high, 85 per cent., and the speed too low, 232 revolutions per minute. With another computation with large assumed friction of the water, we found

$$E = 79 \text{ per cent. hydraulic efficiency,}$$

and

$$n = 228 \text{ revolutions per minute.}$$

These revolutions are much less than those given in Fig. 56, or than has been found in practice. The face angle of the bucket, measured on a drawing of the wheel, is

$$\gamma_1 = 51°,$$

and with this angle we found results agreeing fairly with those found by Dr. Sellers, as communicated to the author, so we have used this value. This is our first opportunity of determining from actual trials whether γ_1 for a finite stream should be the angle made by the face of the bucket with a tangent to the wheel, or the mean angle, and the indication is quite clear that it should be the angle of the face. With the data

$$
\begin{aligned}
&r_1 = 2.625 \text{ ft.} \quad &&\alpha = 19° \quad &&\mu_1 = 0.10 \\
&r_2 = 3.125 \text{ "} \quad &&\gamma_1 = 51° \quad &&\mu_2 = 0.15 \\
&H = 138 \quad \text{"} \quad &&\gamma_2 = 13° \, 17' \quad &&g = 32.16 \text{ ft.,}
\end{aligned}
$$

$2\,r = 38$ in. outside diameter of the tubular part of the shaft.

$t = \frac{3}{4}$ in. thickness of tube of shaft.
$d = 11$ in. outside diameter of solid part of shaft.

We find from equations (15) to (22) ; and the equation preceding (144) :

$M^2 = -0.478 \qquad N^2 = -0.156 \qquad \omega^2 = 642.99.$
$\therefore \omega' = 25.357$ angular velocity per second;
$\therefore N = 242.2$ rev. per minute ;
$E = 0.8108$, or 81 per cent. efficiency;
$v_1 = 22.98$ ft. velocity of entrance into bucket ;
$v_2 = 82.91$ " terminal velocity in the bucket ;
$V = 55.27$ " velocity of quitting the guide;
$V_2 = 19.13$ " quitting velocity in reference to the earth ;
$\theta = 85° 35'$;
$H.P. = 5,500$ horse-power ;
$Q = 433.27$ cu. ft. per second ;
$y_2 = 1.55$ ft.
$y_1 = 1.52$ "
$\omega' r_1 = 66.56$ ft. velocity of inner rim ;
$\omega' r_2 = 79.24$ " " " outer "
$J = 867$ lbs. stress on the tubular part of the shaft.

Several of these results differ perceptibly from those given in Fig. 56. Assuming 250 revolutions per minute, we have

$$\omega = \frac{250 \times 2\pi}{60} = 26.16 \text{ ft. per second.}$$

$\omega r_1 = 68.30$ ft. velocity of initial rim.
$\omega r_2 = 81.81$ " " " terminal rim.

And if $\gamma_1 = 180° - 110° 40' = 69° 20'$ and $\alpha = 19° 00'$, the triangle of velocities gives

$V = 64.31$ ft. the velocity of quitting the guides.
$v_1 = 22.5$ ft., nearly, the initial velocity in the bucket.

And if the ratio of the initial normal section of the bucket to that of the terminal be as 4.275 to 1.25 = 3.42, then

$v_2 = 3.42 \times 22\frac{1}{2} = 76.95$ ft. terminal velocity in the bucket,

$V_2 = 19$ ft. actual velocity of quitting,

and

$\theta = 113°$, direction of V_2.

These figures agree so nearly, almost exactly, with those given in Fig. 56, that we assume that they were obtained in this manner. If the wheel makes 250 revolutions per minute when producing its highest efficiency, under a head of 136 feet, and the other data be as given or assumed, the solution is correct; but otherwise, it is only an approximation, more or less rough. As stated above, the data in Fig. 56 gives about 232 revolutions per minute for best effect, and our computation with $v_1 = 51°$ gives 242; so that if it had 75 per cent. efficiency and gave 5,000 horse-power at 250 revolutions, it ought to have a higher efficiency and greater capacity at a slower speed. If 430 cubic feet were discharged at a velocity of 76.95 feet, through 32 buckets each $1\frac{1}{4}$ inches wide, the depth should be

$$y_2 = \frac{430}{\frac{1}{12} \text{ of } 1.25 \times 32 \times 76.95} = 1.68 \text{ feet.}$$

The actual depth of the six chambers is 1.81 feet; hence the capacity of the wheel should exceed 5,000 horse-power in the ratio 1.81/1.68, giving 5320 horse-power.

This assumes that the buckets are properly made. It appears that the cross-section $a\,d$, Fig. 56, is slightly less than that at $f\,e$, whereas the former ought to be perceptibly larger, since the velocity of the water increases as it goes outward, so that if the section at $a\,d$ is filled, that at $f\,e$ will not be full, and the wheel will be a "pressure" wheel from the initial element to $a\,d$, and one of "free deviation" from $a\,d$ to exit. This being the case, as determined from the plan of the wheel, the correct depth would be found by using the velocity at a, which will be somewhat less than at e, and also the width at a, which is also somewhat less than at e; and both these elements conspire to make the depth y_2 greater than 1.68 as computed. We would modify the design of the wheel by terminating the long arc at, or near, f, and using a shorter radius to, or near, c. Figs. 54 and 54a are suggestive of a better form than that given in this wheel. But retaining the width $e\,f$ and increasing it at $a\,d$ ought to increase the capacity of wheel somewhat without decreasing the efficiency.

We now return to our computation. If the results obtained from equations (15) to (21) do not agree with those found in the wheel, then equations (56) to (62) must be used, since the sections of the buckets are fixed. First, the cross-sections of the initial and terminal sections of the buckets must be inversely as the velocities $v_2/v_1 = 82.91/22.98 = 3.679$.

To determine the ratio of the sections of the stream we assume that the initial section of the stream is the same as that of the bucket, but, as has been shown above, the terminal section of the stream is less than that of the bucket; and to find what the former is—or what the section of the bucket ought to be—requires a knowledge of the capacity of the wheel. This requires an extended analysis, not here given, according to which and to other information, it is for 138 foot head about 5,500 horsepowers; and we will assume this value for the present computation, and test its correctness by the results which follow:

The volume of water flowing through the wheel, in producing 5,500 horse-power, will be

$$Q = \frac{\overset{\text{H.P.}}{5,500} \times \overset{\substack{\text{ft. lb.}\\\text{per sec.}\\\text{per H.P.}}}{550}}{\underset{\substack{\text{per}\\\text{cent.}}}{81} \times \underset{\text{ft.}}{138} \times \underset{\substack{\text{wt.}\\\text{cu. ft.}}}{62.4}} \; 100 = 433 \text{ cu. ft. per sec.}$$

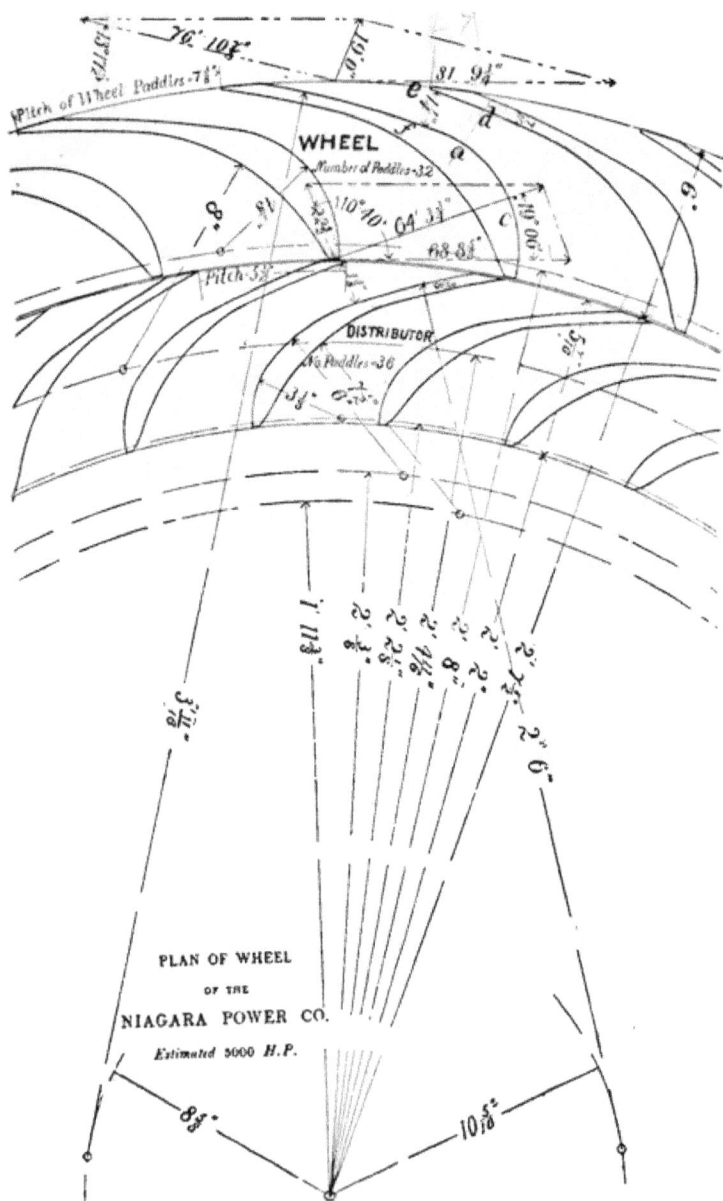

WHEEL
Number of Paddles-32

Pitch of Wheel Paddles-7⅛"

76'.10⅞"

0.6T

81.9¼"

110°10'.04'⅓"

68.5¼"

Pitch-5⅛

DISTRIBUTOR

No Paddles-36

PLAN OF WHEEL

OF THE

NIAGARA POWER CO.

Estimated 5000 H.P.

8⅜"

10⁵⁄₁₆"

FIG. 56.

The terminal velocity being 83 ft. to the nearest entire foot, as found above, the aggregate depth of the six chambers being 1.81 feet, the aggregate width of the 32 buckets should be

$$433 \div (83 \times 1.81) = 2.882 \text{ ft.},$$

and the width of each would be

$$12 \times 2.882 \div 32 = 1.081 \text{ in.},$$

instead of $1\frac{1}{4}$ in. as marked on the plan.

The thickness of the initial end of the partitions between the buckets as measured on the drawing is $\frac{1}{5}$ of an inch each, and for the 32 buckets the aggregate thickness will be $32 \times \dfrac{1}{12 \times 5}$ ft. If, therefore, the buckets were indefinitely narrow, the aggregate thickness of all the partitions being the same, the aggregate normal widths would be

$$2 \pi r_1 \sin 51° - 32 \cdot \frac{1}{12 \times 5} = 12.27 \text{ ft.},$$

and the ratio of the sections in this wheel being as their widths, we have

$$\frac{k_1}{k_2} = \frac{12.27}{2.882} = 4.25.$$

But for buckets of finite width, the ratio is found more accurately by tracing a curved line cutting normally the traces of the buckets, if there were an indefinitely large number. This process gives a ratio of about 4 or a little more. We find that the results do not differ largely for the ratios 4 and 4.25, except for the value of v_1, the initial velocity in the bucket; so we take the ratio 4 as representing more accurately the actual wheel, and as being sufficiently accurate for our purpose, and this differs so much from the inverse ratio of the velocities (3.679) as to make a computation with equations (56) to (62) desirable. With the data

$$K = k_1/k_2 = 4.00, \qquad \mu_1 = 0.10,$$
$$r_1 = 2.675, \qquad \mu_2 = 0.10, \qquad \gamma_1 = 51°,$$
$$r_2 = 3.125, \qquad H = 138 \text{ ft.} \qquad \gamma_2 = 13° 17',$$

we find

$$c = 1/K = 0.25, \qquad a = 1.11203, \qquad b = 3.092,$$
$$A = 0.503, \qquad B = 0.897, \qquad J^2 = -4.221,$$
$$C^2 = -3.9678 \qquad F^2 = 1.602, \qquad G = 2.531.$$

$$\therefore E_{max} = 0.834,$$
$$\omega = 26.45,$$
$$n = 252.5,$$
$$V = 54.54,$$
$$v_2 = 85.29 \text{ ft.,}$$
$$v_1 = 21.32 \text{ "}$$
$$\Gamma_2 = 19.62 \text{ "}$$
$$\omega' r_1 = 69.43 \text{ "}$$
$$\omega' r_2 = 82.65 \text{ "}$$
$$\theta = 88° 53',$$
$$\alpha = 14° 10',$$
$$H. P. = 5.500,$$
$$Q = 433 \text{ cu. ft. per sec.,}$$

Least breadth of bucket = 1.04 in.

No allowance is here made for leakage through the $\frac{1}{16}$-in. clearance of the wheel. There are three such clearances for the escape of water, two at the lower part and one at the upper part of the wheel. From equation (44) it is found that the pressure at entrance into the wheel is

$$p_1 = 7,468 \text{ pounds per sq. foot;}$$

hence, if the coefficient of discharge be 0.80, the volume of discharge will be

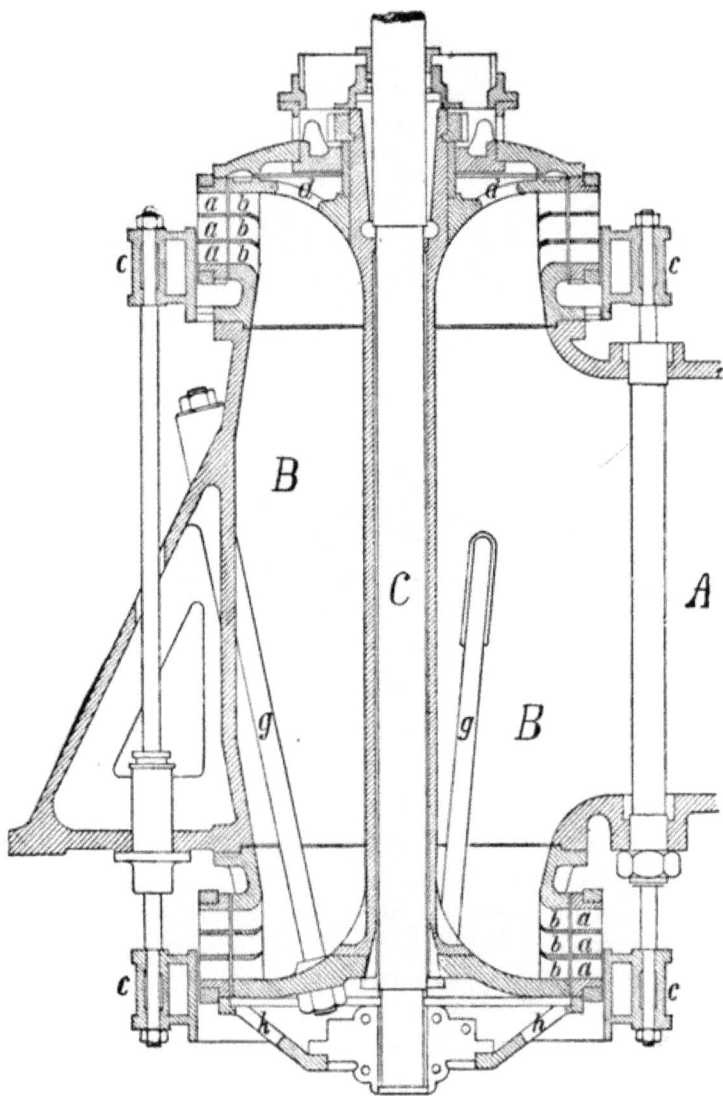

FIG. 57.

$$q = 0.8 \times 2\,\pi\,r_1 \times \frac{3}{12 \times 16}\sqrt{2\,g\,\frac{p_1 - p_2}{62.4}} = 15.4 \text{ cu. ft. per sec.}$$

Hence, the quantity of water passing into the penstock, when running with full gate and at best effect, should be about

$$Q = 433 + 16 = 449 \text{ cu. ft. per sec.}$$

This computed leakage will be

$$\frac{15.4}{448.4} = 0.0344,$$

or 3.44 per cent. of the water delivered to the penstock; or 96.56 per cent. of the water delivered to the penstock passes through the wheel; hence the efficiency of the wheel system referred to the water consumed will be

$$E = 83.4 \times .9656 = 80.53 \text{ per cent.}$$

The terminal angle of the guide (19°) is somewhat larger than given by theory.

We make the following abstract of a statement in regard to a test furnished by Dr. Sellers.

At the time of the test the total head from the surface of the water above the penstock to the centre of the wheel was

$$H = 135.113 \text{ ft.,}$$

and the water delivered to the penstock per minute was

$$Q = 26867 \text{ cu. ft. per min.} = 447.8 \text{ cu. ft. per sec.,}$$

and the theoretical horse-power of the water,

$$H.\,P. = \frac{447.8 \times 135.113 \times 62.3}{550} = 6864.$$

There was an electrical output of 5335 horse-powers; hence the actual efficiency of the wheel and dynamo combined was

$$E' = \frac{5335}{6864} = 0.7785,$$

or, 77.85 per cent. ; and if the dynamo yielded 97 per cent. as
guaranteed by the makers, then the efficiency of the wheel system, including friction and leakage, would be

$$E = \frac{77.85}{0.97} = 80.26 \text{ per cent.,}$$

and the power delivered at the upper end of the shaft would be

$$H. P. = \frac{5335}{0.97} = 5.500.$$

The head during this test was less than that assumed in the
computation, but if 1.4 ft. for the head due to the velocity in
the penstock, be added, the effective head will be 136.5 ft.,
which is only 1.5 ft. less than the effective head assumed. This
difference will not affect the efficiency, but would affect the com-
parative speed. The speed was not measured, but was regulated
for about 245 to 250 revolutions per minute, and the experimental efficiency and power involve an assumption ; and the theoretical computation is founded on the supposition that the wheel
is a pressure wheel throughout ; so that it cannot be said whether
a more exact analysis would agree more nearly with a test experiment, if the data were precisely the same and the quantities
directly measured. As they stand, the two results—theory and
experiment—agree remarkably well. The indication is—the resistances are less than those assumed, the leakage greater than
that computed, and the hydraulic efficiency greater than 85 per
cent. of the power of the water passing through the wheel.

The volume of water may now be recomputed, and will be

$$Q = \frac{5,500 \times 550}{62.4 \times 138 \times 0.8053} = 436,$$

which is 3 cu. ft. more than that before found, which difference is chiefly due to the difference in the efficiency used

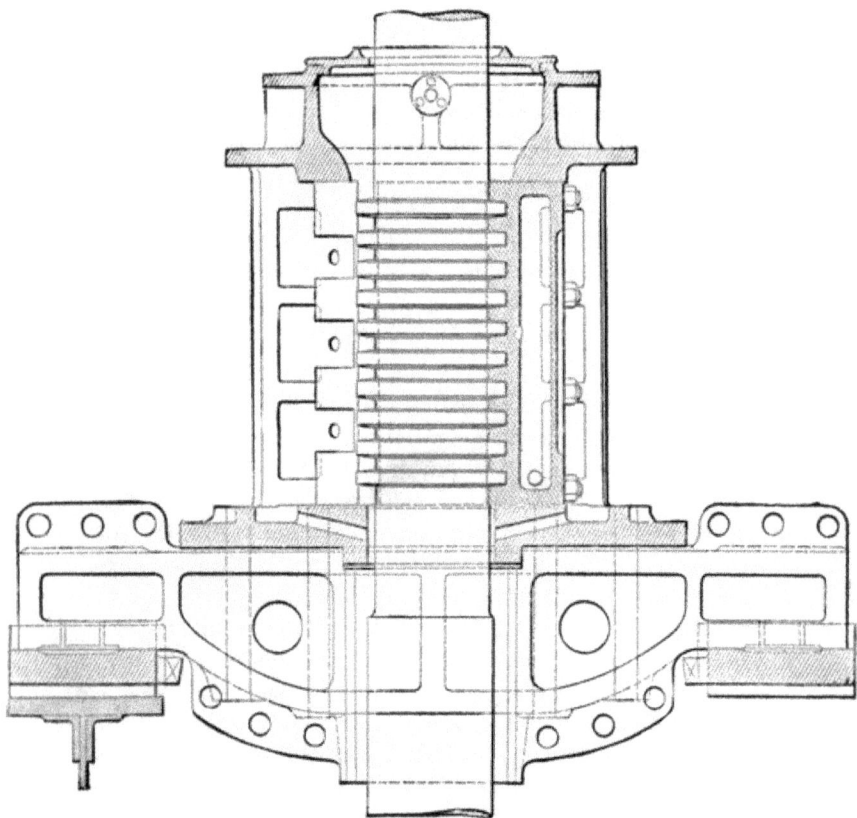

Fig. 59.

in the computation. Using this result, the terminal normal width of the bucket will be :

$$\text{Width} = \frac{436 \times 12}{1.51 \times 85.29 \times 32} = 1.04 \text{ in.,}$$

which is 0.04 of an inch less than that found by the former computation, which is due to the larger terminal velocity now found. This again emphasizes the remark previously made in regard to increasing the capacity of the wheel.

According to our computation, the velocity of the water in the penstock will be

$$v = \frac{433}{\frac{1}{4}\,\pi\,(7\frac{1}{2})^2} = 9.8 \text{ ft.}$$

The velocity as it enters the case will be . . . . . 12.2 "
   "   "   in the case just before entering the distributers will be . . . . 28.5 "
   "   "   entering the distributers will be . . . 30.4 "
   "   "   quitting   "   "   " ".   . . . 55.3 "
   "   "   entering the wheel relative to the bucket will be . . . . 21.3 "
   "   "   quitting the bucket will be . . . . . 85.3 "
   "   "   "   "   wheel in reference to the earth will be . . . . 19.6 "

The main part of the shaft is a tube of steel rolled and without longitudinal riveted seam, 38 in. outside diameter and $\frac{3}{4}$ in. thick. There are two solid parts joining the tubular parts, as shown in Fig. 59, which form journals for the support of the shaft and wheel, and are 11 in. in diameter, one of which is shown in Fig. 58. The moment of stress is given in Article 49, and is

$$12\,P\,a = \frac{63,000\ H.P.}{n} \text{ inch pounds.}$$

For the resistance, let

r be the mean radius of the tubular shaft in inches;

t, the thickness of the tube ;

J, the modulus of torsional shear ;

then for a thin tube

$$2 \pi r . t$$

will be the sectional area of the tube,

$$2 \pi r t . J$$

will be the resisting force of the tube, and

$$2 \pi r t J . r$$

will be the moment of resistance ;

$$\therefore 2 \pi r^2 t J = \frac{63,000 \ H.P.}{n} ;$$

$$\therefore J = \frac{31,500 \ H.P.}{\pi r^2 t n},$$

which in this case becomes,

$$J = \frac{31500 \times 5,500}{3.14 \times 18\frac{5}{8}^2 \times \frac{3}{4} \times 252} = 867 \ \text{pounds.}$$

The torsional stress on the solid part will be given by the equation

$$\frac{63,000 \ H.P.}{n} = \frac{1}{2} \pi J R^3,$$

in which R is the external radius of the solid part and is $5\frac{1}{2}$ in. :

$$\therefore J = 8 \ \frac{126,000 \ H.P.}{\pi n . 11^3} = 5,260 \ \text{pounds.}$$

The resistance to shear of steel or iron in large masses is not well known. If homogeneous, theory indicates that it will be $\frac{4}{5}$ of the tenacity of the material, and experiments indicate that the shearing resistance is nearly the same as that of the tenacity. The tenacity of mild steel is 65,000 pounds and upward per square inch ; hence its shearing strength ought to be 50,000 pounds at least, according to which the solid part will be strained to about $\frac{1}{10}$ of its ultimate strength when running steadily and delivering 5500 horse-powers, which is no more than ought to be allowed for safety, considering that in starting and stopping and for variations of loads, the stress may be considerably increased. The stress on the tubular part is small compared with that on the solid part—less than $\frac{1}{5}$ as great. If the shaft be a uniform tube $\frac{3}{4}$ in. thick, 19 in. radius, 140 ft. long, and if the modulus of elasticity to shear be 10,000,000 pounds, then will the amount of torsion, when running steady at 252 revolutions, delivering 5500 horse-powers, be

$$\varphi = \frac{63,000 \times 5,500 \times 140 \times 12}{10,000,000 \times 2 \ \pi \ r \ t \ . \ r^2 \ . \ n} = 0.07364,$$

which is the arc for radius unity.

The number of degrees will be

$$\frac{\varphi}{3.14} \times 180 = 4° \ 13'.$$

Fig. 59 shows the penstock, shaft, and relative position of the wheel. They are supported by heavy cast-iron beams resting on the solid rock.

Pressure due to deflecting a stream.

87. The pressure resulting from the deflection of stream of water may be determined as follows : Let a particle whose mass is m enter a stationary tube with a velocity, v, and follow the tube to its exit. Let the tube be frictionless, then will the velocity be v in reference to the tube from entrance to exit. Let the tube be the arc of a circle with O as the centre. The centrifugal force will be radially outward and equal to

$$m \ \frac{v^2}{r} \ ;$$

which will be the pressure against the outside of the tube, and may be represented in magnitude and direction by the line $A \ B$

on the radius $O\,A$ prolonged. The centripetal force will be of equal magnitude, and will be the reaction of the tube acting upon the particle toward the centre O. Assume that the particle fills the tube for a distance $d\,s$; then if

k be the cross-section of the tube;

δ the weight of unity of volume;

θ the angle $D\,O\,A$, Fig. 60;

r the radius $O\,A$;

φ the centrifugal force; then

$$d\,s = r\,d\,\theta;$$

$$d\,m = \frac{\delta\,k\,d\,s}{g};$$

$$\therefore d\,\varphi = \frac{\delta\,k\,v^2}{g\,r}\,r\,d\,\theta.$$

Resolve this force into two forces, $B\,C$ parallel to $O\,x$ and $A\,C$ perpendicular thereto, then will the sum of all the pressures parallel to $O\,x$ due to the particle in passing from D to A be

$$X_\theta = \frac{\delta\,k\,v^2}{g}\,\int_0^{} \sin\theta\,d\,\theta,$$

$$= \frac{\delta\,k\,v^2}{g}\,(1 - \cos\theta),$$

where X_θ represents the x-component of the pressure.

Instead of a single particle moving in the tube let a stream of liquid flow through the tube, and let

M be the mass flowing into the tube per second, then

$$M = \frac{\delta\,k\,v}{g};$$

hence the constant pressure in the direction $O\,x$ will be

$$X_\theta = M\,v\,(1 - \cos\theta). \tag{147}$$

If

$$\theta = \tfrac{1}{2}\,\pi,$$

then

$$X_{\frac{1}{2}\pi} = M\,v, \tag{148}$$

or the resultant pressure equals the momentum of the jet per second.

If

$$\theta = \pi$$

$$X_\pi = 2\,M\,v, \tag{149}$$

or the resultant pressure will be twice the momentum of the jet.

88. Now assume that the tube *moves* in the direction $O\,x$ with an uniform velocity u, and that M is the mass of the stream per section k entering the tube, the relative velocity will be

$$v - u,$$

and the resultant pressure in that direction for an arc subtending θ,

$$M\,(v - u)\,(1 - \cos\,\theta), \tag{150}$$

and if θ be 90° the work per second will be

$$P\,u = M\,(v - u)\,u. \tag{151}$$

This will be a maximum for $u = \frac{1}{2}\,v$, for which

$$P\,u = \frac{1}{4}\,M\,v^2, \tag{152}$$

or one half the energy of the stream will be utilized in doing work, the other half remaining in the stream. This result may be deduced directly, thus: The relative velocity at the entrance D being $v - u = \frac{1}{2}\,v$, and the tube being frictionless, the velocity at exit will also be $\frac{1}{2}\,v$; hence at E the water will have a component of $\frac{1}{2}\,v$ parallel to $O\,x$ and another of $\frac{1}{2}\,v$ perpendicular to the same, the resultant of which will be

$$\sqrt{(\tfrac{1}{2}\,v)^2 + (\tfrac{1}{2}\,v)^2} = \sqrt{\tfrac{1}{2}} \cdot v,$$

and the energy will be $\frac{1}{4}\,M\,v^2$; the remaining half being transformed into work. If the mass be that which passes a fixed point, represented by M' having the section k, then will the mass entering the tube be

$$M'\left(1 - \frac{u}{v}\right), \tag{153}$$

and the work will be

$$P u = M' \frac{(v - u)^2}{v} u, \qquad (154)$$

which will be a maximum for $u = \frac{1}{3} v$;

$$\therefore P u = \frac{4}{27} M' v^2. \qquad (155)$$

But this is referring the energy to a mass, a part of which is not delivered to the wheel, a method which is not ordinarily practised. Moreover, in practice floats or buckets are usually made to follow each other so closely that the stream of the section of the buckets is assumed to be all delivered to the wheel, and equation (151) is applicable. This, then, is the case of an impulse wheel, or wheel of free deviation, in which the direction of the terminal element is $\gamma_2 = 90°$. By assuming that the water immediately in front of the floats in a "paddle" or undershot wheel, Fig. 61, forms a guide for the water which subsequently follows, the analysis for the work of the plane float undershot wheel results in equation (151), and hence such a wheel cannot utilize more than one half the energy of the stream : and as there are inevitable wastes from the escaping water through the clearances, the practical efficiency will be still less, and experiments have shown it to be about

$$0.30 \ M \ v^2. \qquad (156)$$

and except for wheels well made, mechanically, it will be still smaller. We have thus brought this wheel within the analysis of the turbine of free deviation.

89. It has been stated in Article 6 that equation (16) gives $\omega_1 = 0$, for $\gamma_2 = 90°$. If $\gamma_2 = 0$ be substituted in (15) it becomes

$$E = L M^2 . r_2^2 \omega^2, \qquad . \qquad (157)$$

which increases indefinitely with the square of the speed of the wheel, and hence has no maximum. Also if $\alpha = 0$ and $\gamma_1 = 180°$, M and

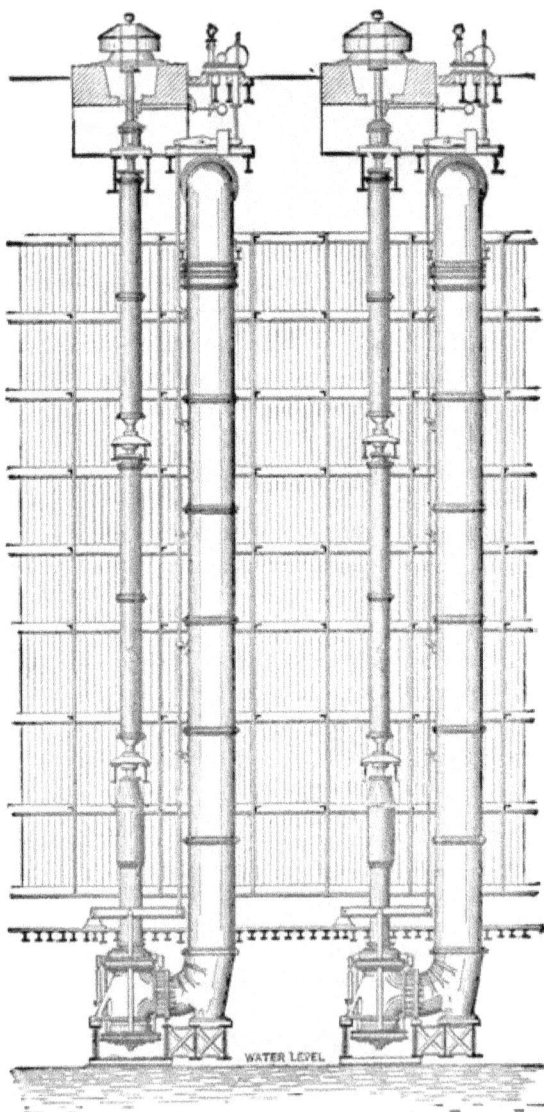

WATER LEVEL

Fig. 59.

N contain ambiguous terms of the form $\frac{0}{0}$, some of which when evaluated give infinity; so that the general equation seems to fail for these particular values; because there is no maximum for this case. But for the wheel of "free deviation" there is a maximum for this condition as shown by the following analysis:

For this case the velocity V will depend directly on the head, and making $p_1 = p_a$, and $h_1 = H$ in equation (4), we have

$$(1 + \mu_1)\ V^2 = 2\ g\ H, \tag{158}$$

as has already been given in equation (73), page 42.

From the triangle $A\ B\ C$, Fig. 1. find e, in terms of V and ω, giving

$$v_1^2 = V^2 + r_1^2\ \omega^2 - 2\ V \cos \alpha \, . \, r_1\ \omega, \tag{159}$$

as also given in equation (94); and equation (9) gives

$$(1 + \mu_2)\ v_2^2 = v_1^2 + (r_2^2 - r_1^2)\ \omega^2 ; \tag{160}$$

equation (11) becomes

$$V_2^2 = v_2^2 + r_2^2\ \omega^2 ; \tag{161}$$

and (12) and (13) give

$$U = \delta\ Q\ H - \frac{\delta\ Q}{2\ g}\ [\ V_2^2 + \mu_1\ V^2 + \mu_2\ v_2^2],$$

$$= \frac{\delta\ Q}{g}\ [-\ r_2^2\ \omega^2 + V \cos \alpha \, \cdot \, r_1\ \omega] \tag{162}$$

$$\therefore E = \frac{U}{\delta\ Q\ H} = \frac{1}{g\ H}\ [-\ r_2^2\ \omega^2 + V \cos \alpha \, \cdot \, r_1\ \omega] ; \tag{163}$$

which is a maximum for

$$r_2\ \omega = \tfrac{1}{2}\ V\ \frac{r_1}{r_2} \cos \alpha, \tag{164}$$

in which $r_2\ \omega$ will be the velocity of the outer rim $= V''$ (say) and if the buckets are very narrow, or if they move in a right line, as in the preceding article, or practically in a right line, we may consider

$$r_1 = r_2;$$

then

$$V'' = \tfrac{1}{2} V \cos \alpha, \tag{165}$$

or the velocity of the bucket will be one half the velocity of the component of the velocity of the water as it issues from the supply chamber in the direction of motion of the wheel. If $\alpha = 0$, the velocity of the bucket should be one half that of the jet ; and for this case we have

$$V = \sqrt{\frac{2 g H}{1 + \mu_1}}. \tag{166}$$

$$c_1 = \tfrac{1}{2} V. \tag{167}$$

$$(1 + \mu_2) c_2^2 = \tfrac{1}{4} V^2. \tag{168}$$

$$V_2^2 = \tfrac{1}{2} V^2 \tag{169}$$

$$C = \frac{\delta \, Q}{2 \, g} \cdot \tfrac{1}{2} V^2 = \tfrac{1}{2} \delta \, Q \, H. \tag{170}$$

$$E = \tfrac{1}{2}. \tag{171}$$

Equations (162) and (163) show that the work and efficiency are independent of the frictional resistances of the water, but (166) and (168) show that V and c_2 are diminished by such resistances.

If γ_2 be finite and the wheel one of " free deviation," we have from equations (158), (159), (11), (12), and (13), after making $d E \div d \omega = 0$,

$$0 = \sqrt{1 + \mu_2} \, (V r_1 \cos \alpha - 2 \, r_2^2 \, \omega) +$$

$$r_2 \cos \gamma_2 \, \frac{2 V^2 - 3 V r_1 \cos \alpha \cdot \omega + 4 \, r_2^2 \, \omega^2}{\sqrt{V^2 - 2 V r_1 \cos \alpha \cdot \omega + r_2^2 \, \omega^2}}.$$

This produces a complete equation of the fourth degree, and the complete solution will be more lengthy than for the pressure wheel, as given in equations (16) and (60).

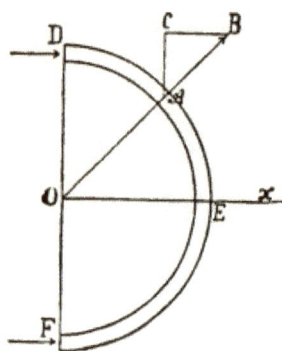

FIG. 60.

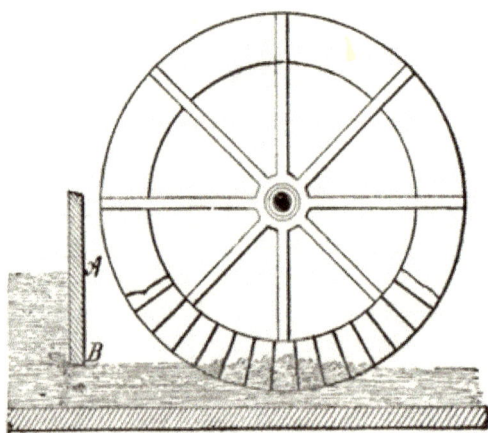

FIG. 61.

90. *Cascade Wheel.*—The James Leffel & Co. have designed two wheels to be established at Ward, Col., to work under a head of 730 feet. One is to be 38 inches diameter, to be driven by one nozzle, producing about 25 horse-power at about 552 revolutions per minute; the other, 50 inches diameter, fed by a nozzle a little larger than 1⅓ inches, will have a capacity of about 140 horse-power. Some wheels of this design are now working under high heads. Fig. 62 shows one of these wheels with pulley. In actual running the wheel is inclosed in a case, which was removed in order to show the construction of the wheel.

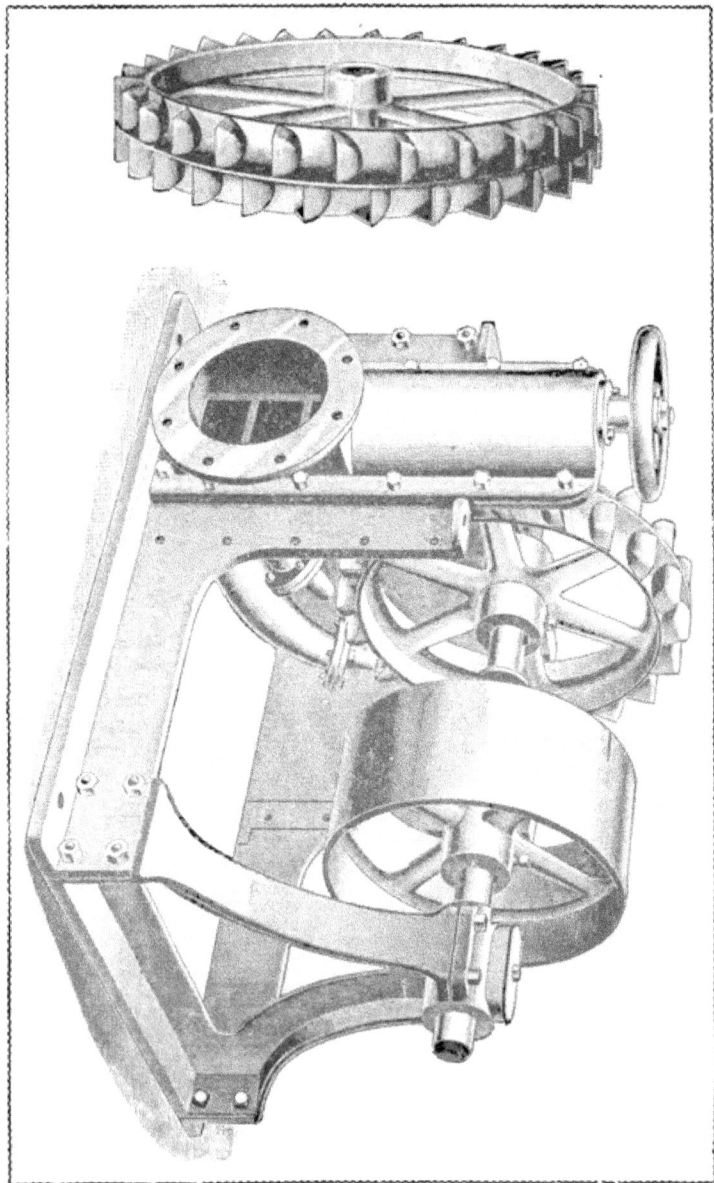

Fig. 92.

INDEX.

	PAGE
Angular velocity	10, 19, 29
Barker's mill	48, 64
Boott turbine	89
Boyden diffuser	89
Bryden turbine	92
Bucket, depth of	89
" form of	17, 86, 123
" initial angle of	10, 14, 122, 133
" terminal angle of	122
Buckets, number of	25, 122
Centrifugal force	5, 41
Class-room exercise	117
Coefficient of effect	21
Collins turbine	94
Comparison of inward and outward flow wheels	35
" " various makes of turbines	130
Crown, width of	121
Crowns	27
" depth between	72
" form of	125
" parallel	66
Designing	3, 27, 117
Diameter of wheel	120
Diffuser, Bryden	80
Direction of quitting water	9, 18, 84
Efficiency	25
" effect of γ on	15
" equation for	7, 28, 29
" maximum	8
" " for inflow	35
" " " outflow	35
" wheel of free deviation	45
Energy imparted to wheel	13
" lost by impact	26
" " in escaping water	26
" loss of due to quitting velocity	35
Escaping water, volume of	73
Exercises	25, 31, 41, 43, 46, 49, 70, 90, 117
Fall, power of	117

	PAGE
Form of bucket	17, 86, 123
" " crowns	125
Fourneyron turbine	37, 93
Francis and Thomson's vortex wheel	37
Free deviation	18, 43, 44, 147
" surface, form of	45
Friction along buckets	26
" head lost by	4
Frictional resistance	5, 13
Frictionless wheel, work done in	39
" " efficiency of	39
" " path of water for	40
General solution of pressure turbine	3
Guides, terminal angle of	10, 11, 35, 122
Haenel turbine	97
Head due to pressure and velocity	4
" lost by friction	4
" total	4
" virtual	3
Hercules turbine	107
Humphrey wheel	102
Hurdy-gurdy wheel	114
Inflow wheel, efficiency of	15, 26
" " maximum efficiency of	35
Initial angle of bucket	10, 14, 122, 133
" velocity	4
Inward flow	9
" " ratio of radii for	25
Jet propeller	50
Jonval turbine	37
Leakage	137
Maximum efficiency	7
Moment of stress	85, 141
Momentum, moment of	56
Niagara wheel	130
Notation	1
Number of buckets	25
Outflow wheel efficiency compared with inflow	35
Outward flow	9

	PAGE
Outward flow efficiency of	26
" " ratio of radii for	25
Parallel crowns	66
" flow turbine	9, 93
Path of water	30, 85, 128
Pelton wheel	111
Poncelot wheel	114
Pressure at entrance to bucket	23
" " exit from bucket	6
" due to deflecting stream	143
" in wheel	22, 82
" theoretical at entrance to bucket	4
" turbines, general solution of	3
Quitting angle, value of	9, 18, 84
Radii, ratio of	25
Rankine wheel	37, 65
Relation between γ_2 and ω^1	10
Revolutions for best efficiency	47
Risdon wheel	101
Scottish and Whitelaw turbine	50
Segmental feed	95, 96
Shaft, diameter of	85
Stress, moment of	85, 141
Submerged wheels	23, 24
Swain turbines	103
Tables, effect of different values of γ_1	14
" " " γ_1 on velocities	15
' for revolutions and H.P. for best efficiency	47
Tangential wheels	99
Terminal angle, effect of large	78
" " of bucket	122
" " " guide	10, 11
" velocity	4
Tests, Boott turbine	90
" Bryden turbine	92
" Collins turbine	94
" Haenel turbine	97
" Swain turbine	103
" Tremont turbine	68, 69
" Victor turbine	110
Total head	4
Tremont turbine	67
Turbine, Boott	89
" Boyden	92

	PAGE
Turbine, Collins	94
" Fourneyron	37, 93
" " cut	4
" " triple, cut	10
" Guideless, cut	24
" Haenel	97
" Hercules	107
" Inflow, cut	12
" Jet, cut	26
" Jonval	37, 93
" mixed flow, cut	20
" outflow, cut	2
" parallel flow	93
" " " cut	14, 16
" Scottish and Whitelaw	50
" Swain	103
" Tangential, cut	22
" Tremont	67
" Victor	109
Values of a	11
" " $a + \gamma_1$	10
" " θ	18
" " μ_1 and μ_2	13, 34, 70, 71, 78, 138
" " ω^1	19
Velocity along rotating tube	41
" at exit	6
" of entering water	8, 28
" " initial rim	12
" " quitting water	9, 28
" " terminal rim	12
" " wheel for best effect	10
Velocities, initial	4
" terminal	4
Victor turbine	109
Virtual head	3
Volume escaping water	73
" of water flowing through wheel	136
Vortex wheel, Francis and Thomson's	37
Wheel of free deviation	41, 147
Work, equation for	7
" done by centrifugal force	5
" " " falling weight	5
" " upon wheel	6
" lost by friction	32

SHORT-TITLE CATALOGUE

OF THE

PUBLICATIONS

OF

JOHN WILEY & SONS,

NEW YORK.

LONDON: CHAPMAN & HALL, LIMITED.

ARRANGED UNDER SUBJECTS.

Descriptive circulars sent on application.
Books marked with an asterisk are sold at *net* prices only.
All books are bound in cloth unless otherwise stated.

AGRICULTURE.

CATTLE FEEDING—DISEASES OF ANIMALS—GARDENING, ETC.

Armsby's Manual of Cattle Feeding...............12mo,	$1	75
Downing's Fruit and Fruit Trees.......................8vo,	5	00
Kemp's Landscape Gardening.... ....12mo,	2	50
Stockbridge's Rocks and Soils.... ....8vo,	2	50
Lloyd's Science of Agriculture..........................8vo,	4	00
Loudon's Gardening for Ladies. (Downing.)...........12mo,	1	50
Steel's Treatise on the Diseases of the Ox.................8vo,	6	00
" Treatise on the Diseases of the Dog................8vo,	3	50
Grotenfelt's The Principles of Modern Dairy Practice. (Woll.)		
12mo,	2	00

ARCHITECTURE.

BUILDING—CARPENTRY—STAIRS, ETC.

Berg's Buildings and Structures of American Railroads.....4to,	7	50
Birkmire's Architectural Iron and Steel..................8vo,	3	50
" Skeleton Construction in Buildings...........8vo,	3	00

Birkmire's Compound Riveted Girders.............. ......8vo, $2 00

      "   American Theatres—Planning and Construction.8vo, 3 00

Carpenter's Heating and Ventilating of Buildings.........8vo, 3 00

Freitag's Architectural Engineering.... .. ..8vo, 2 50

Kidder's Architect and Builder's Pocket-book.....Morocco flap, 4 00

Hatfield's American House Carpenter.................. .. 8vo, 5 00

    "   Transverse Strains...........................8vo, 5 00

Monckton's Stair Building—Wood, Iron, and Stone.......4to, 4 00

Gerhard's Sanitary House Inspection....................16mo, 1 00

Downing and Wightwick's Hints to Architects......... ..8vo, 2 00

    "   Cottages........ ....8vo, 2 50

Holly's Carpenter and Joiner.. ....18mo, 75

Worcester's Small Hospitals- ·Establishment and Maintenance,

    including Atkinson's Suggestions for Hospital Archi-

    tecture... ....12mo, 1 25

The World's Columbian Exposition of 1893............ .. 4to, 2 50

ARMY, NAVY, Etc.

Military Engineering—Ordnance—Port Charges, Etc.

Cooke's Naval Ordnance 8vo, $12 50

Metcalfe's Ordnance and Gunnery..........12mo, with Atlas, 5 00

Ingalls's Handbook of Problems in Direct Fire...........8vo, 4 00

    "   Ballistic Tables................................8vo, 1 50

Bucknill's Submarine Mines and Torpedoes..............8vo, 4 00

Todd and Whall's Practical Seamanship 8vo, 7 50

Mahan's Advanced Guard............ ....18mo, 1 50

    "   Permanent Fortifications. (Mercur.).8vo, half morocco, 7 50

Wheeler's Siege Operations...............................8vo, 2 00

Woodhull's Notes on Military Hygiene.........12mo, morocco, 2 50

Dietz's Soldier's First Aid...................12mo, morocco, 1 25

Young's Simple Elements of Navigation..12mo, morocco flaps, 2 50

Reed's Signal Service....................................... 50

Phelps's Practical Marine Surveying......................8vo, 2 50

Very's Navies of the World...............8vo, half morocco, 3 50

Bourne's Screw Propellers................................4to, 5 00

Hunter's Port Charges....................8vo, half morocco, $13 00
* Dredge's Modern French Artillery...........4to, half morocco, 20 00
   "   Record of the Transportation Exhibits Building,
     World's Columbian Exposition of 1893..4to, half morocco, 15 00
Mercur's Elements of the Art of War..................8vo, 4 00
   "   Attack of Fortified Places....................12mo, 2 00
Chase's Screw Propellers....................................8vo, 3 00
Winthrop's Abridgment of Military Law..............12mo, 2 50
De Brack's Cavalry Outpost Duties. (Carr)....18mo, morocco, 2 00
Cronkhite's Gunnery for Non-com. Officers.....18mo, morocco, 2 00
Dyer's Light Artillery. 12mo, 3 00
Sharpe's Subsisting Armies..........................18mo, 1 25
   "   "   "   ....18mo, morocco, 1 50
Powell's Army Officer's Examiner....................12mo, 4 00
Hoff's Naval Tactics................................8vo, 1 50
Bruff's Ordnance and Gunnery..........................8vo, 6 00

ASSAYING.

SMELTING—ORE DRESSING—ALLOYS, ETC.

Furman's Practical Assaying............................8vo, 3 00
Wilson's Cyanide Processes..........................12mo, 1 50
Fletcher's Quant. Assaying with the Blowpipe..12mo, morocco, 1 50
Ricketts's Assaying and Assay Schemes... ....8vo, 3 00
Mitchell's Practical Assaying. (Crookes.)..............8vo, 10 00
Thurston's Alloys, Brasses, and Bronzes..............8vo, 2 50
Kunhardt's Ore Dressing............................8vo, 1 50
O'Driscoll's Treatment of Gold Ores.....................8vo, 2 00

ASTRONOMY.

PRACTICAL, THEORETICAL, AND DESCRIPTIVE.

Michie and Harlow's Practical Astronomy...............8vo, 3 00
White's Theoretical and Descriptive Astronomy..........12mo, 2 00
Doolittle's Practical Astronomy........................8vo, 4 00
Craig's Azimuth......................................4to, 3 50
Gore's Elements of Geodesy............................8vo. 2 50

3

BOTANY.

GARDENING FOR LADIES, ETC.

Westermaier's General Botany. (Schneider.)............8vo,	$2 00
Thomé's Structural Botany..............................18mo,	2 25
Baldwin's Orchids of New England.................. .....8vo,	1 50
Loudon's Gardening for Ladies. (Downing.)............12mo,	1 50

BRIDGES, ROOFS, Etc.

CANTILEVER—HIGHWAY—SUSPENSION.

Boller's Highway Bridges................................8vo,	2 00
* " The Thames River Bridge.................4to, paper,	5 00
Burr's Stresses in Bridges.... ....8vo,	3 50
Merriman & Jacoby's Text-book of Roofs and Bridges. Part I., Stresses..... ....8vo,	2 50
Merriman & Jacoby's Text-book of Roofs and Bridges. Part II., Graphic Statics...............................8vo.	2 50
Merriman & Jacoby's Text-book of Roofs and Bridges. Part III., Bridge Design................................8vo.	5 00
Merriman & Jacoby's Text-book of Roofs and Bridges. Part IV., Continuous, Draw, Cantilever, Suspension, and Arched Bridges......................(In preparation).	
Crehore's Mechanics of the Girder.......................8vo,	5 00
Du Bois's Strains in Framed Structures...................4to,	10 00
Greene's Roof Trusses...................................8vo,	1 25
" Bridge Trusses...............................8vo,	2 50
" Arches in Wood, etc...........................8vo,	2 50
Waddell's Iron Highway Bridges....... ....8vo,	4 00
Wood's Construction of Bridges and Roofs..............8vo,	2 00
Foster's Wooden Trestle Bridges.........................4to,	5 00
* Morison's The Memphis Bridge.................Oblong 4to,	10 00
Johnson's Modern Framed Structures....................4to,	10 00

CHEMISTRY.

QUALITATIVE—QUANTITATIVE—ORGANIC—INORGANIC, ETC.

Fresenius's Qualitative Chemical Analysis. (Johnson.)....8vo,	4 00
" Quantitative Chemical Analysis. (Allen.).......8vo,	6 00
" " " " (Bolton.).....8vo.	1 50

4

Crafts's Qualitative Analysis.   (Schaeffer.).................12mo, $1 50
Perkins's Qualitative Analysis... ....12mo,   1 00
Thorpe's Quantitative Chemical Analysis................18mo,   1 50
Classen's Analysis by Electrolysis.   (Herrick.)............8vo,   3 00
Stockbridge's Rocks and Soils............................8vo,   2 50
O'Brine's Laboratory Guide to Chemical Analysis. 8vo,   2 00
Mixter's Elementary Text-book of Chemistry............12mo,   1 50
Wulling's Inorganic Phar. and Med. Chemistry..........12mo,   2 00
Mandel's Bio-chemical Laboratory......................12mo,   1 50
Austen's Notes for Chemical Students..................12mo,
Schimpf's Volumetric Analysis........................12mo,   2 50
Hammarsten's Physiological Chemistry (Mandel.)..........8vo,   4 00
Miller's Chemical Physics..............................8vo,   2 00
Pinner's Organic Chemistry.   (Austen.)................12mo,   1 50
Kolbe's Inorganic Chemistry...........................12mo,   1 50
Ricketts and Russell's Notes on Inorganic Chemistry (Non-
        metallic)........ .... Oblong 8vo, morocco,   75
Drechsel's Chemical Reactions. (Merrill.)...............12mo,   1 25
Adriance's Laboratory Calculations.....................12mo,   1 25
Troilius's Chemistry of Iron............................8vo,   2 00
Allen's Tables for Iron Analysis........... ....8vo,
Nichols's Water Supply (Chemical and Sanitary).........8vo,   2 50
Mason's   "   "   "   "   " 8vo,   5 00
Spencer's Sugar Manufacturer's Handbook.12mo, morocco flaps,   2 00
Wiechmann's Sugar Analysis........... ....8vo,   2 50
        "   Chemical Lecture Notes................12mo,   3 00

DRAWING.

ELEMENTARY—GEOMETRICAL—TOPOGRAPHICAL.

Hill's Shades and Shadows and Perspective....(*In preparation*)
Mahan's Industrial Drawing. (Thompson.).......2 vols., 8vo,   3 50
MacCord's Kinematics..................................8vo,   5 00
    "   Mechanical Drawing.........................8vo,   4 00
    "   Descriptive Geometry............. ....8vo,   3 00
Reed's Topographical Drawing.   (II. A.)................4to,   5 00
Smith's Topographical Drawing. (Macmillan.)...........8vo,   2 50
Warren's Free-hand Drawing 12mo,   1 00

Warren's Drafting Instruments......................12mo, $1 25
"   Projection Drawing......................12mo, 1 50
"   Linear Perspective......................12mo, 1 00
"   Plane Problems..........................12mo, 1 25
"   Primary Geometry........................12mo,   75
"   Descriptive Geometry..............2 vols., 8vo, 3 50
"   Problems and Theorems.....................8vo, 2 50
"   Machine Construction............2 vols., 8vo, 7 50
"   Stereotomy—Stone Cutting...................8vo, 2 50
"   Higher Linear Perspective 8vo, 3 50
"   Shades and Shadows.........................8vo, 3 00
Whelpley's Letter Engraving.....................12mo, 2 00

ELECTRICITY AND MAGNETISM.
ILLUMINATION—BATTERIES—PHYSICS.

* Dredge's Electric Illuminations....2 vols., 4to, half morocco, 25 00
   "   "   "   Vol. II..................4to, 7 50
Niaudet's Electric Batteries. (Fishback.)............12mo, 2 50
Anthony and Brackett's Text-book of Physics...........8vo, 4 00
Cosmic Law of Thermal Repulsion...................18mo,   75
Thurston's Stationary Steam Engines for Electric Lighting Pur-
   poses..12mo, 1 50
Michie's Wave Motion Relating to Sound and Light........8vo, 4 00
Barker's Deep-sea Soundings..........................8vo, 2 00
Holman's Precision of Measurements...................8vo, 2 00
Tillman's Heat..8vo, 1 50
Gilbert's De-magnete. (Mottelay.)....................8vo, 2 50
Benjamin's Voltaic Cell..............................8vo, 3 00
Reagan's Steam and Electrical Locomotives............12mo 2 00

ENGINEERING.
CIVIL—MECHANICAL—SANITARY, ETC.

* Trautwine's Cross-section..........................Sheet,   25
*   "   Civil Engineer's Pocket-book...12mo, mor. flaps, 5 00
*   "   Excavations and Embankments............8vo, 2 00
*   "   Laying Out Curves...........12mo, morocco, 2 50
Hudson's Excavation Tables. Vol. II..................8vo, 1 00

Searles's Field Engineering...............12mo, morocco flaps, $3 00
" Railroad Spiral ....12mo, morocco flaps, 1 50
Godwin's Railroad Engineer's Field-book.12mo, pocket-bk. form, 2 50
Butts's Engineer's Field-book.................12mo, morocco, 2 50
Gore's Elements of Goodesy....... ....8vo, 2 50
Wellington's Location of Railways... ....8vo, 5 00
*Dredge's Penn. Railroad Construction, etc... Folio, half mor., 20 00
Smith's Cable Tramways....................................4to, 2 50
" Wire Manufacture and Uses......................4to, 3 00
Mahan's Civil Engineering. (Wood.)...................8vo, 5 00
Wheeler's Civil Engineering...........................8vo, 4 00
Mosely's Mechanical Engineering. (Mahan.)....... ......8vo, 5 00
Johnson's Theory and Practice of Surveying..............8vo, 4 00
" Stadia Reduction Diagram..Sheet, 22½ × 28½ inches, 50
* Drinker's Tunnelling.....................4to, half morocco, 25 00
Eissler's Explosives—Nitroglycerine and Dynamite........8vo, 4 00
Foster's Wooden Trestle Bridges..........................4to, 5 00
Ruffner's Non-tidal Rivers............... ....8vo, 1 25
Greene's Roof Trusses 8vo, 1 25
" Bridge Trusses................................8vo, 2 50
" Arches in Wood, etc............. ...8vo, 2 50
Church's Mechanics of Engineering—Solids and Fluids....8vo, 6 00
" Notes and Examples in Mechanics..............8vo, 2 00
Howe's Retaining Walls (New Edition.)................12mo, 1 25
Wegmann's Construction of Masonry Dams...............4to, 5 00
Thurston's Materials of Construction....................8vo, 5 00
Baker's Masonry Construction......................... 8vo, 5 00
" Surveying Instruments........................12mo, 3 00
Warren's Stereotomy—Stone Cutting....................8vo, 2 50
Nichols's Water Supply (Chemical and Sanitary)......... 8vo, 2 50
Mason's " " " " " 8vo, 5 00
Gerhard's Sanitary House Inspection....................16mo, 1 00
Kirkwood's Lead Pipe for Service Pipe...................8vo, 1 50
Wolff's Windmill as a Prime Mover......................8vo, 3 00
Howard's Transition Curve Field-book.....12mo, morocco flap, 1 50
Crandall's The Transition Curve ....12mo, morocco, 1 50
7

Crandall's Earthwork Tables 8vo, $1 50
Patton's Civil Engineering..............................8vo, 7 50
" Foundations......................................8vo, 5 00
Carpenter's Experimental Engineering 8vo, 6 00
Webb's Engineering Instruments.............12mo, morocco, 1 00
Black's U. S. Public Works.............................4to, 5 00
Merriman and Brook's Handbook for Surveyors....12mo, mor., 2 00
Merriman's Retaining Walls and Masonry Dams.........8vo, 2 00
" Geodetic Surveying.........................8vo, 2 00
Kiersted's Sewage Disposal..... ....12mo, 1 25
Siebert and Biggin's Modern Stone Cutting and Masonry...8vo, 1 50
Kent's Mechanical Engineer's Pocket-book.....12mo, morocco, 5 00

HYDRAULICS.

WATER-WHEELS—WINDMILLS—SERVICE PIPE—DRAINAGE, ETC.

Weisbach's Hydraulics. (Du Bois.)......................8vo, 5 00
Merriman's Treatise on Hydraulics.................... ...8vo, 4 00
Ganguillet & Kutter's Flow of Water. (Hering & Trautwine).8vo, 4 00
Nichols's Water Supply (Chemical and Sanitary)..........8vo, 2 50
Wolff's Windmill as a Prime Mover......................8vo, 3 00
Ferrel's Treatise on the Winds, Cyclones, and Tornadoes...8vo, 4 00
Kirkwood's Lead Pipe for Service Pipe..................8vo, 1 50
Ruffner's Improvement for Non-tidal Rivers..............8vo, 1 25
Wilson's Irrigation Engineering.... ....8vo, 4 00
Bovey's Treatise on Hydraulics.........................8vo, 4 00
Wegmann's Water Supply of the City of New York 4to, 10 00
Hazen's Filtration of Public Water Supply................8vo, 2 00
Mason's Water Supply—Chemical and Sanitary...........8vo, 5 00
Wood's Theory of Turbines.... ....8vo, 2 50

MANUFACTURES.

ANILINE—BOILERS—EXPLOSIVES—IRON—SUGAR—WATCHES—
WOOLLENS, ETC.

Metcalfe's Cost of Manufactures.......................8vo, 5 00
Metcalf's Steel (Manual for Steel Users)................12mo, 2 00
Allen's Tables for Iron Analysis........................8vo,

West's American Foundry Practice....... ....12mo, $2 50
"   Moulder's Text-book 12mo,   2 50
Spencer's Sugar Manufacturer's Handbook....12mo, mor. flap,   2 00
Wiechmann's Sugar Analysis........................8vo,   2 50
Beaumont's Woollen and Worsted Manufacture........12mo,   1 50
* Reisig's Guide to Piece Dyeing......................8vo, 25 00
Eissler's Explosives, Nitroglycerine and Dynamite........8vo,   4 00
Reimann's Aniline Colors. (Crookes.)................8vo,   2 50
Ford's Boiler Making for Boiler Makers................18mo,   1 00
Thurston's Manual of Steam Boilers.................... 8vo,   5 00
Booth's Clock and Watch Maker's Manual.............12mo,   2 00
Holly's Saw Filing... ....18mo,   75
Svedelius's Handbook for Charcoal Burners............12mo,   1 50
The Lathe and Its Uses............................ .. 8vo,   6 00
Woodbury's Fire Protection of Mills....................8vo,   2 50
Bolland's The Iron Founder...........................12mo,   2 50
"   "   "   "   Supplement................12mo,   2 50
"   Encyclopædia of Founding Terms............12mo,   3 00
Bouvier's Handbook on Oil Painting...................12mo,   2 00
Steven's House Painting.............................18mo,   75

MATERIALS OF ENGINEERING.

STRENGTH—ELASTICITY—RESISTANCE, ETC.

Thurston's Materials of Engineering...............3 vols., 8vo,   8 00
        Vol. I., Non-metallic................ ....8vo,   2 00
        Vol. II., Iron and Steel.......................... 8vo,   3 50
        Vol. III., Alloys, Brasses, and Bronzes.............8vo,   2 50
Thurston's Materials of Construction............... ......8vo,   5 00
Baker's Masonry Construction..........................8vo,   5 00
Lanza's Applied Mechanics.............................8vo,   7 50
"   Strength of Wooden Columns ....8vo, paper,   50
Wood's Resistance of Materials........................8vo,   2 00
Weyrauch's Strength of Iron and Steel. (Du Bois.).......8vo,   1 50
Burr's Elasticity and Resistance of Materials............8vo,   5 00
Merriman's Mechanics of Materials.....................8vo,   4 00
Church's Mechanic's of Engineering—Solids and Fluids.....8vo,   6 00
9

Beardslee and Kent's Strength of Wrought Iron 8vo, $1 50

Hatfield's Transverse Strains............................8vo, 5 00

Du Bois's Strains in Framed Structures..................4to, 10 00

Merrill's Stones for Building and Decoration............8vo, 5 00

Bovey's Strength of Materials...........................8vo, 7 50

Spalding's Roads and Pavements....................12mo, 2 00

Rockwell's Roads and Pavements in France............12mo, 1 25

Byrne's Highway Construction......... ....8vo, 5 00

Patton's Treatise on Foundations.......................8vo, 5 00

MATHEMATICS.

CALCULUS—GEOMETRY—TRIGONOMETRY, ETC.

Rice and Johnson's Differential Calculus.................8vo, 3 50

" Abridgment of Differential Calculus....8vo, 1 50

" Differential and Integral Calculus.

2 vols. in 1, 12mo, 2 50

Johnson's Integral Calculus...........................12mo, 1 50

" Curve Tracing.............................12mo, 1 00

" Differential Equations—Ordinary and Partial.....8vo, 3 50

" Least Squares...............................12mo, 1 50

Craig's Linear Differential Equations................. ...8vo, 5 00

Merriman and Woodward's Higher Mathematics........ ...8vo,

Bass's Differential Calculus...........................12mo,

Halsted's Synthetic Geometry...........................8vo, 1 50

" Elements of Geometry...................8vo, 1 75

Chapman's Theory of Equations........................12mo, 1 50

Merriman's Method of Least Squares 8vo, 2 00

Compton's Logarithmic Computations..................12mo, 1 50

Davis's Introduction to the Logic of Algebra.............8vo, 1 50

Warren's Primary Geometry...........................12mo, 75

" Plane Problems........................... 12mo, 1 25

" Descriptive Geometry................ 2 vols., 8vo, 3 50

" Problems and Theorems......................8vo, 2 50

" Higher Linear Perspective...................8vo, 3 50

" Free-hand Drawing................, ...12mo, 1 00

" Drafting Instruments...................... 12mo, 1 25

Warren's Projection Drawing..........................12mo, $1 50
    "   Linear Perspective...........................12mo, 1 00
    "   Plane Problems..............................12mo, 1 25
Searles's Elements of Geometry. 8vo, 1 50
Brigg's Plane Analytical Geometry......................12mo, 1 00
Wood's Co-ordinate Geometry............................8vo, 2 00
    "   Trigonometry................................12mo, 1 00
Mahan's Descriptive Geometry (Stone Cutting)...........8vo, 1 50
Woolf's Descriptive Geometry.....................Royal 8vo, 3 00
Ludlow's Trigonometry with Tables. (Bass.)............8vo, 3 00
    "   Logarithmic and Other Tables. (Bass.)........8vo, 2 00
Baker's Elliptic Functions..............................8vo, 1 50
Parker's Quadrature of the Circle. 8vo, 2 50
Totten's Metrology......................................8vo, 2 50
Ballard's Pyramid Problem...............................8vo, 1 50
Barnard's Pyramid Problem...............................8vo, 1 50

MECHANICS—MACHINERY.

TEXT-BOOKS AND PRACTICAL WORKS.

Dana's Elementary Mechanics.........................12mo, 1 50
Wood's   "   " 12mo, 1 25
    "   "   "   Supplement and Key.......... 1 25
    "   Analytical Mechanics..........................8vo, 3 00
Michie's Analytical Mechanics...........................8vo, 4 00
Merriman's Mechanics of Materials......................8vo, 4 00
Church's Mechanics of Engineering......................8vo, 6 00
    "   Notes and Examples in Mechanics..............8vo, 2 00
Mosely's Mechanical Engineering. (Mahan.).............8vo, 5 00
Weisbach's Mechanics of Engineering. Vol. III., Part I.,
    Sec. I. (Klein.)...................................8vo, 5 00
Weisbach's Mechanics of Engineering. Vol. III., Part I.
    Sec. II. (Klein.)..................................8vo, 5 00
Weisbach's Hydraulics and Hydraulic Motors. (Du Bois.)..8vo, 5 00
    "   Steam Engines. (Du Bois.).....................8vo, 5 00
Lanza's Applied Mechanics..............................8vo, 7 50
11

Crehore's Mechanics of the Girder.........................8vo, $5 00
MacCord's Kinematics.................................8vo, 5 00
Thurston's Friction and Lost Work..... ....8vo, 3 00
    "   The Animal as a Machine 12mo, 1 00
Hall's Car Lubrication.................................12mo, 1 00
Warren's Machine Construction..................2 vols., 8vo, 7 50
Chordal's Letters to Mechanics....................... 12mo, 2 00
The Lathe and Its Uses.............................. .. 8vo, 6 00
Cromwell's Toothed Gearing............................12mo, 1 50
    "   Belts and Pulleys.........................12mo, 1 50
Du Bois's Mechanics. Vol. I., Kinematics 8vo, 3 50
    "   "   Vol. II., Statics.. ....8vo, 4 00
    "   "   Vol. III., Kinetics....... ....,8vo, 3 50
Dredge's Trans. Exhibits Building, World Exposition,
                            4to, half morocco, 15 00
Flather's Dynamometers...............................12mo, 2 00
    "   Rope Driving...............................12mo, 2 00
Richards's Compressed Air............................12mo, 1 50
Smith's Press-working of Metals......................8vo, 3 00
Holly's Saw Filing 18mo, 75
Fitzgerald's Boston Machinist..................... ....18mo, 1 00
Baldwin's Steam Heating for Buildings.................12mo, 2 50
Metcalfe's Cost of Manufactures........................8vo, 5 00
Benjamin's Wrinkles and Recipes.......................12mo, 2 00
Dingey's Machinery Pattern Making 12mo, 2 00

METALLURGY.

Iron—Gold—Silver—Alloys, Etc.

Egleston's Metallurgy of Silver........................ 8vo, 7 50
    "   Gold and Mercury............................8vo, 7 50
    "   Weights and Measures, Tables..............18mo, 75
    "   Catalogue of Minerals.......................8vo, 2 50
O'Driscoll's Treatment of Gold Ores....................8vo, 2 00
* Kerl's Metallurgy—Copper and Iron...................8vo, 15 00
*   "   "   Steel, Fuel, etc...................8vo, 15 00

Thurston's Iron and Steel................................8vo, $3 50
    "   Alloys......................................8vo,   2 50
Troilius's Chemistry of Iron.....................8vo,   2 00
Kunhardt's Ore Dressing in Europe......................8vo,   1 50
Weyrauch's Strength of Iron and Steel. (Du Bois.)........**8vo,   1 50**
Beardslee and Kent's Strength of Wrought Iron...........**8vo,   1 50**
Compton's First Lessons in Metal Working.............12mo,   1 50
West's American Foundry Practice....................12mo,   2 50
   "   Moulder's Text-book...........................12mo,   2 50

MINERALOGY AND MINING.

MINE ACCIDENTS—VENTILATION—ORE DRESSING, ETC.

Dana's Descriptive Mineralogy. (E. S.).....8vo, half morocco,   12 50
  "   Mineralogy and Petrography. (J. D.)...........12mo,   2 00
  "   Text-book of Mineralogy. (E. S.)................8vo,   3 50
  "   Minerals and How to Study Them. (E. S.)......12mo,   1 50
  "   American Localities of Minerals...................8vo,   1 00
Brush and Dana's Determinative Mineralogy.. ....8vo,   3 50
Rosenbusch's Microscopical Physiography of Minerals and
     Rocks. (Iddings.)..................................8vo,   5 00
Hussak's Rock-forming Minerals. (Smith.)..............8vo,   2 00
Williams's Lithology.......................................8vo,   3 00
Chester's Catalogue of Minerals...... ....8vo,   1 25
    "   Dictionary of the Names of Minerals.............8vo,   3 00
Egleston's Catalogue of Minerals and Synonyms..........8vo,   2 50
Goodyear's Coal Mines of the Western Coast..........12mo,   2 50
Kunhardt's Ore Dressing in Europe 8vo,   1 50
Sawyer's Accidents in Mines...........................8vo,   7 00
Wilson's Mine Ventilation............. ....16mo,   1 25
Boyd's Resources of South Western Virginia.............8vo,   3 00
   "   Map of South Western Virginia.....Pocket-book form,   2 00
Stockbridge's Rocks and Soils...........................8vo,   2 50
Eissler's Explosives—Nitroglycerine and Dynamite........8vo,   4 00
13

*Drinker's Tunnelling, Explosives, Compounds, and Rock Drills.
4to, half morocco, $25 00
Beard's Ventilation of Mines.. ....12mo, 2 50
Ihlseng's Manual of Mining.. ....8vo, 4 00

STEAM AND ELECTRICAL ENGINES, BOILERS, Etc.

STATIONARY—MARINE—LOCOMOTIVE—GAS ENGINES, ETC.

Weisbach's Steam Engine. (Du Bois.)...................8vo, 5 00
Thurston's Engine and Boiler Trials.....................8vo, 5 00
    "   Philosophy of the Steam Engine............12mo, 75
    "   Stationary Steam Engines..................12mo, 1 50
    "   Boiler Explosion.... ....12mo, 1 50
    "   Steam-boiler Construction and Operation.......8vo,
    "   Reflection on the Motive Power of Heat. (Carnot.)
12mo, 2 00
Thurston's Manual of the Steam Engine. Part I., Structure
    and Theory.......................................8vo, 7 50
Thurston's Manual of the Steam Engine. Part II., Design,
    Construction, and Operation......................8vo, 7 50
2 parts, 12 00
Röntgen's Thermodynamics. (Du Bois.)................8vo, 5 00
Peabody's Thermodynamics of the Steam Engine......... 8vo, 5 00
    "   Valve Gears for the Steam-Engine.............8vo, 2 50
    "   Tables of Saturated Steam....................8vo, 1 00
Wood's Thermodynamics, Heat Motors, etc..............8vo, 4 00
Pupin and Osterberg's Thermodynamics................12mo, 1 25
Kneass's Practice and Theory of the Injector 8vo, 1 50
Reagan's Steam and Electrical Locomotives....... .....12mo, 2 00
Meyer's Modern Locomotive Construction.................4to, 10 00
Whitham's Steam-engine Design 8vo, 6 00
    "   Constructive Steam Engineering...............8vo, 10 00
Hemenway's Indicator Practice........................12mo, 2 00
Pray's Twenty Years with the Indicator............Royal 8vo, 2 50
Spangler's Valve Gears.................................8vo, 2 50
* Maw's Marine Engines.................Folio, half morocco, 18 00
Trowbridge's Stationary Steam Engines ....4to, boards, 2 50

Ford's Boiler Making for Boiler Makers..................18mo, $1 00
Wilson's Steam Boilers. (Flather.)....................12mo, 2 50
Baldwin's Steam Heating for Buildings.................12mo, 2 50
Hoadley's Warm-blast Furnace..........................8vo, 1 50
Sinclair's Locomotive Running.........................12mo, 2 00
Clerk's Gas Engine...................... .. ..12mo,

TABLES, WEIGHTS, AND MEASURES.

For Engineers, Mechanics, Actuaries—Metric Tables, Etc.

Crandall's Railway and Earthwork Tables..... ....8vo, 1 50
Johnson's Stadia and Earthwork Tables.......... ......8vo, 1 25
Bixby's Graphical Computing Tables......... ....Sheet, 25
Compton's Logarithms.................................12mo, 1 50
Ludlow's Logarithmic and Other Tables. (Bass.)......12mo, 2 00
Thurston's Conversion Tables...................... ... ...8vo, 1 00
Egleston's Weights and Measures......................18mo, 75
Totten's Metrology.......................... ....8vo, 2 50
Fisher's Table of Cubic Yards....................Cardboard, 25
Hudson's Excavation Tables. Vol. II..... ....8vo, 1 00

VENTILATION.

Steam Heating—House Inspection—Mine Ventilation.

Beard's Ventilation of Mines 12mo, 2 50
Baldwin's Steam Heating..............................12mo, 2 50
Reid's Ventilation of American Dwellings 12mo, 1 50
Mott's The Air We Breathe, and Ventilation 16mo, 1 00
Gerhard's Sanitary House Inspection ....Square 16mo, 1 00
Wilson's Mine Ventilation.............................16mo, 1 25
Carpenter's Heating and Ventilating of Buildings.........8vo, 3 00

MISCELLANEOUS PUBLICATIONS.

Alcott's Gems, Sentiment, Language..............Gilt edges, 5 00
Bailey's The New Tale of a Tub.................... ....8vo, 75
Ballard's Solution of the Pyramid Problem..............8vo, 1 50
Barnard's The Metrological System of the Great Pyramid..8vo, 1 50
15

* Wiley's Yosemite, Alaska, and Yellowstone 4to, $3 00
Emmon's Geological Guide-book of the Rocky Mountains..8vo, 1 50
Ferrel's Treatise on the Winds..........................8vo, 4 00
Perkins's Cornell University.....................Oblong 4to, 1 50
Ricketts's History of Rensselaer Polytechnic Institute.....8vo, 3 00
Mott's The Fallacy of the Present Theory of Sound..Sq. 16mo, 1 00
Rotherham's The New Testament Critically Emphathized.
                                                       12mo, 1 50
Totten's An Important Question in Metrology............8vo, 2 50
Whitehouse's Lake Mœris.............................Paper, 25

HEBREW AND CHALDEE TEXT-BOOKS.

For Schools and Theological Seminaries.

Gesenius's Hebrew and Chaldee Lexicon to Old Testament.
    (Tregelles.).... .... Small 4to, half morocco, 5 00
Green's Grammar of the Hebrew Language (New Edition).8vo, 3 00
   "   Elementary Hebrew Grammar... ....12mo, 1 25
   "   Hebrew Chrestomathy...........................8vo, 2 00
Letteris's Hebrew Bible (Massoretic Notes in English).
                                       8vo, arabesque, 2 25
Luzzato's Grammar of the Biblical Chaldaic Language and the
    Talmud Babli Idioms...........................12mo, 1 50

MEDICAL.

Bull's Maternal Management in Health and Disease......12mo, 1 00
Mott's Composition, Digestibility, and Nutritive Value of Food.
                                  Large mounted chart, 1 25
Steel's Treatise on the Diseases of the Ox.... ....8vo, 6 00
   "   Treatise on the Diseases of the Dog...............8vo, 3 50
Worcester's Small Hospitals—Establishment and Maintenance,
    including Atkinson's Suggestions for Hospital Archi-
    tecture...12mo, 1 25
Hammarsten's Physiological Chemistry. (Mandel.).......8vo, 4 00

16

www.ingramcontent.com/pod-product-compliance
Lightning Source LLC
Chambersburg PA
CBHW021518210326

41599CB00012B/1306